Probabilistic Modelling

Probabilistic Modelling

Isi Mitrani
University of Newcastle

PUBLISHED BY THE PRESS SYNDICATE OF THE UNIVERSITY OF CAMBRIDGE
The Pitt Building, Trumpington Street, Cambridge CB2 1RP, United Kingdom

CAMBRIDGE UNIVERSITY PRESS
The Edinburgh Building, Cambridge CB2 2RU, United Kingdom
40 West 20th Street, New York, NY 10011-4211, USA
10 Stamford Road, Oakleigh, Melbourne 3166, Australia

© Cambridge University Press 1998

This book is in copyright. Subject to statutory exception
and to the provisions of relevant collective licensing agreements,
no reproduction of any part may take place without
the written permission of Cambridge University Press

First published 1998

Printed in the United Kingdom at the University Press, Cambridge

A catalogue record for this book is available from the British Library

Library of Congress Cataloguing in Publication data

Mitrani, I.
Probabilistic modelling / Isi Mitrani.
p. cm.
Includes bibliographical references and index.
ISBN 0-521-58511-2 (hc)
1. Probabilities. 2. Mathematical models. I. Title.
QA273.M595 1997
003'.85–dc21 97-26099 CIP

ISBN 0 521 58511 2 hardback
ISBN 0 521 58530 9 paperback

Contents

Preface		*page* ix
1	**Introduction to probability theory**	1
1.1	Sample points, events and probabilities	1
1.1.1	An algebra of events	3
1.1.2	Probabilities	4
1.1.3	Conditional probability	7
1.2	Random variables	10
1.2.1	Distribution functions	11
1.2.2	Probability density functions	13
1.2.3	Joint distributions	15
1.2.4	Largest and smallest	18
1.3	Expectation and other moments	20
1.3.1	Properties of expectation	23
1.3.2	Variance and covariance	24
1.4	Bernoulli trials and related random variables	28
1.4.1	The geometric distribution	28
1.4.2	The binomial distribution	31
1.5	Sums, transforms and limits	36
1.5.1	Generating functions	37
1.5.2	Laplace transforms and characteristic functions	39
1.5.3	The normal distribution and central limit theorem	41
1.6	Literature	46
2	**Arrivals and services**	47
2.1	Renewal processes	47
2.1.1	Forward and backward renewal times	50
2.1.2	The 'paradox' of residual life	53
2.2	The exponential distribution and its properties	55

2.2.1	First and last	56
2.3	The Poisson process	59
2.3.1	Properties of the Poisson process	61
2.3.2	Arrivals during a random interval	67
2.4	Application: ALOHA and CSMA	69
2.5	Literature	73
3	**Queueing systems: average performance**	**74**
3.1	Little's theorem and applications	75
3.1.1	Utilization and response time laws	77
3.2	The single-class $M/G/1$ queue	81
3.2.1	Busy periods	86
3.2.2	The $M/M/1$ queue	88
3.3	Different scheduling policies	92
3.3.1	The Processor-Sharing policy	94
3.3.2	Symmetric policies and multiclass queues	96
3.3.3	Which policy is better?	98
3.4	Priority scheduling	99
3.4.1	Preemptive-resume priorities	104
3.4.2	Averaging and lumping job types	105
3.5	Optimal scheduling policies	107
3.5.1	A conservation law for waiting times	108
3.5.2	The c/ρ rule	110
3.5.3	Optimization with preemptions	112
3.5.4	Characterization of achievable performance	116
3.6	Literature	119
4	**Queueing networks**	**122**
4.1	Open networks	124
4.1.1	Traffic equations	125
4.1.2	Performance measures	128
4.2	Closed networks	136
4.2.1	Mean value analysis	138
4.3	Multiclass networks	146
4.3.1	Closed multiclass networks	150
4.4	Literature	154
5	**Markov chains and processes**	**156**
5.1	Markov chains	157
5.1.1	Steady state	161
5.2	A Markov chain embedded in the $M/G/1$ model	168
5.3	Markov processes	172

5.3.1	Transient behaviour	178
5.3.2	Steady state	180
5.4	First passages and absorptions	184
5.5	Birth-and-Death queueing models	186
5.5.1	The M/M/n model	188
5.5.2	The M/M/n/n and M/M/∞ systems	192
5.5.3	The M/M/1/N queue	194
5.5.4	The M/M/n/·/K model	195
5.6	Literature	199
6	**Queues in Markovian environments**	**201**
6.1	Quasi-Birth-and-Death models	202
6.1.1	The MMPP/M/1 queue	204
6.1.2	A multiserver queue with breakdowns and repairs	205
6.1.3	Manufacturing blocking	207
6.1.4	Phase-type distributions	208
6.2	Solution methods	210
6.2.1	Spectral expansion method	211
6.2.2	Matrix-geometric solution	215
6.3	Generalizations	216
6.4	Literature	219
Index		221

Preface

The designers and users of complex systems have an interest in knowing how those systems behave under different conditions. This is true in all engineering domains, from transport and manufacturing to computing and communications. It is necessary to have a clear understanding, both qualitative and quantitative, of the factors that influence the performance and reliability of a system. Such understanding may be obtained by constructing and analysing mathematical models. The purpose of this book is to provide the necessary background, methods and techniques.

A model is inevitably an approximation of reality: a number of simplifying assumptions are usually made. However, that need not diminish the value of the insights that can be gained. A mathematical model can capture all the essential features of a system, display underlying trends and provide quantitative relations between input parameters and performance characteristics. Moreover, analysis is cheap, whereas experimentation is expensive. A few simple calculations carried out on the back of an envelope can often yield as much information as hours of observations or simulations.

The systems in which we are interested are subjected to demands of random character. The processes that take place in response to those demands are also random. Accordingly, the modelling tools that are needed to study such systems come from the domains of probability theory, stochastic processes and queueing theory.

Probabilities, random variables and distributions are introduced in chapter 1. Much of the theory concerned with arrivals, services, queues, scheduling policies, optimization and networks can be developed without any more sophisticated mathematical aparatus. This is done in chapters 2, 3 and 4, where the emphasis is on average performance. Chapter 5 deals with the important topic of Markov chains and processes, and their applications. Finally, chapter 6 is devoted to a particular class of models—queues in Markovian environments—and the methods available for their solution.

Throughout the text, an effort has been made to make the material easily accessible. There is much emphasis on explaining ideas and providing intuition along with the formal derivations and proofs. Some of the more difficult results are, in fact, stated without proofs. Generality is sometimes sacrificed to clarity. For instance, the mean value treatment of closed queueing networks was chosen in preference to an approach based on the product–form solution because it is simpler, albeit somewhat less powerful.

The book is intended for operations research and computer science undergraduates and postgraduates, and for practitioners in the field. Some mathematical background is assumed, including first-year calculus. Readers familiar with probability may wish to skip some, or all, of chapter 1.

I would like to thank David Tranah, of Cambridge University Press, for his help in the preparation of the manuscript.

1

Introduction to probability theory

Unpredictability and non-determinism are all around us. The future behaviour of any system—from an elementary particle to a complex organism—may follow a number of possible paths. Some of these paths may be more likely than others, but none is absolutely certain. Such unpredictable behaviour, and the phenomena that cause it, are usually described as 'random'. Whether randomness is in the nature of reality, or is the result of imperfect knowledge, is a philosophical question which need not concern us here. More important is to learn how to deal with randomness, how to quantify it and take it into account, so as to be able to plan and make rational choices in the face of uncertainty.

The theory of probabilities was developed with this object in view. Its domain of applications, which was originally confined mainly to various games of chance, now extends over most scientific and engineering disciplines.

This chapter is intended as a self-contained introduction; it describes all the concepts and results of probability theory that will be used in the rest of the book. Examples and exercises are included. However, it is impossible to provide a thorough coverage of a major branch of mathematics in one chapter. The reader is therefore assumed to have encountered at least some of this material already.

1.1 Sample points, events and probabilities

We start by introducing the notion of a 'random experiment'. Any action, or sequence of actions, which may have more than one possible outcome, can be considered as a random experiment. The set of all possible outcomes is usually denoted by Ω and is called the 'sample space' of the experiment. The individual outcomes, or elements of Ω, are called 'sample points'.

The sample space may be a finite or an infinite set. In the latter case, it may be enumerable (i.e. the outcomes can be numbered $0, 1, \ldots$), or it may be non-enumerable (e.g. Ω can be an interval on the real line).

Examples

1. A race takes place between n horses, with the sole object of determining the winner. There are n possible outcomes, so the sample space can be identified with the set $\Omega = \{1, 2, \ldots, n\}$.

2. In the same race, the finishing position of every horse is of interest and is recorded. The possible outcomes are now the $n!$ permutations of the integers $\{1, 2, \ldots, n\}$: $\Omega = \{(1, 2, \ldots, n), \ldots, (n, n-1, \ldots, 1)\}$ (assuming that two horses cannot arrive at the finishing line at exactly the same time).

3. A certain task is carried out by an unreliable machine which may break down before completion. If that happens, the machine is repaired and the task is restarted from the beginning. The experiment ends when the task is successfully completed; the only output produced is the number of times that the task was attempted. The sample space here is the set of all positive integers: $\Omega = \{1, 2, \ldots\}$.

4. For the same machine, the experiment consists of measuring, with infinite accuracy, the period of time between one repair and the next breakdown. The sample points now are the positive real numbers: $\Omega = \{x : x \in \mathcal{R}^+\}$.

* * *

The next important concept is that of an 'event'. Intuitively, we associate the occurrence of an event with certain outcomes of the experiment. For instance, in example 1, the event 'an even-numbered horse wins the race' is associated with the sample points $\{2, 4, 6, \ldots, n - (n \bmod 2)\}$. In example 3, the event 'the task is run unsuccessfully at least k times' is represented by the sample points $\{k+1, k+2, \ldots\}$. In general, an event is defined as a subset of the sample space Ω. Given such an event, A, the two statements 'A occurs' and 'the outcome of the experiment is one of the points in A' have the same meaning.

1.1.1 An algebra of events

The usual operations on sets—complement, union and intersection—have simple interpretations in terms of occurrence of events. If A is an event, then the complement of A with respect to Ω (i.e. those points in Ω which are not in A) occurs when A does not, and vice versa. Clearly, that complement should also be an event. It is denoted by $\overline{A}$, or $\neg A$, or A^c. If A and B are two events, then the union of A and B (the sample points which belong to either A, or B, or both) is an event which occurs when either A occurs, or B occurs, or both. That event is denoted by $A + B$, or $A \cup B$. Similarly, the intersection of A and B (the sample points which belong to both A and B) is an event which occurs when both A and B occur. It is denoted by AB, or $A \cap B$, or A, B.

There is considerable freedom in deciding which subsets of Ω are to be called 'events' and which are not. It is necessary, however, that the definition should be such that the above operations on events can be carried out. More precisely, the set of all events, $\mathcal{A}$, must satisfy the following three axioms.

E1: The entire sample space, Ω, is an event (this event occurs no matter what the outcome of the experiment).

E2: If A is an event, then $\overline{A}$ is also an event.

E3: If $\{A_1, A_2, \ldots\}$ is any countable set of events, then the union $A = \bigcup_{i=1}^{\infty} A_i$ is also an event.

From E1 and E2 it follows that the empty set, $\emptyset = \overline{\Omega}$, is an event (that event can never occur). From E2 and E3 it follows that if $\{B_1, B_2, \ldots\}$ is a countable set of events, then the intersection

$$B = \bigcap_{i=1}^{\infty} B_i = \overline{\bigcup_{i=1}^{\infty} \overline{B_i}}, \qquad (1.1)$$

is also an event.

The 'countable set' mentioned in E3 may of course be finite:

If $A_1, A_2, \ldots, A_n$ are events, then the union $A = \bigcup_{i=1}^{n} A_i$ is also an event.

Similarly, (1.1) applies to finite intersections.

In set theory, a family which satisfies E1–E3 is called a σ-algebra (or a σ-field, or a Borel field). Thus the set of all events, $\mathcal{A}$, must be a σ-algebra. At one extreme, $\mathcal{A}$ could consist of Ω and $\emptyset$ only; at the other, $\mathcal{A}$ could contain every subset of Ω.

Two events are said to be 'disjoint' or 'mutually exclusive' if they cannot occur together, i.e. if their intersection is empty. More than two events are disjoint if every pair of events among them are disjoint. A set of events $\{A_1, A_2, \ldots\}$ is said to be 'complete', or to be a 'partition of Ω', if (i) those events are mutually exclusive and (ii) their union is Ω. In other words, no matter what the outcome of the experiment, one and only one of those events occurs.

To illustrate these definitions, consider example 3, where a task is given to an unreliable machine to be carried out. Here we can define $\mathcal{A}$ as the set of all subsets of Ω. Two disjoint events are, for instance, $A = \{1, 2, 3\}$ (the task is completed in no more than three runs) and $B = \{5, 6\}$ (it takes five or six runs). However, if we include, say, event $C = \{6, 7, \ldots\}$ (the task needs at least six runs to complete) then the three events A, B, C are not disjoint because B and C are not. The events A and C, together with $D = \{4, 5\}$, form a partition of Ω.

1.1.2 Probabilities

Having defined the events that may occur as a result of an experiment, it is desirable to measure the relative likelihoods of those occurrences. This is done by assigning to each event, A, a number, called the 'probability' of that event and denoted by $P(A)$. By convention, these numbers are normalized so that the probability of an event which is certain to occur is 1 and the probability of an event which cannot possibly occur is 0. The probabilities of all events are in the (closed) interval $[0, 1]$. Moreover, since the probability is, in some sense, a measure of the event, it should have the additive property of measures: just as the area of the union of non-intersecting regions is equal to the sum of their areas, so the probability of the union of disjoint events is equal to the sum of their probabilities.

Thus, probability is a function, P, defined over the set of all events, whose values are real numbers. That function satisfies the following three axioms.

P1: $0 \leq P(A) \leq 1$ for all $A \in \mathcal{A}$.
P2: $P(\Omega) = 1$.

1.1 Sample points, events and probabilities

P3: If $\{A_1, A_2, \ldots\}$ is a countable (finite or infinite) set of *disjoint* events, then $P[\bigcup_{i=1}^{\infty} A_i] = \sum_{i=1}^{\infty} P(A_i)$.

Note that Ω is not necessarily the only event which has a probability of 1. For instance, consider an experiment where a true die is tossed infinitely many times. We shall see later that the probability of the event 'a 6 will appear at least once' is 1. Yet that event is not equal to Ω, because there are outcomes for which it does not occur. In general, if A is an event whose probability is 1, then A is said to occur 'almost certainly'.

An immediate consequence of P2 and P3 is that, if $\{A_1, A_2, \ldots\}$ is a (finite or infinite) partition of Ω, then

$$\sum_{i=1}^{\infty} P(A_i) = 1 . \tag{1.2}$$

In particular, for every event A, $P(\overline{A}) = 1 - P(A)$. Hence, the probability of the empty event is zero: $P(\emptyset) = 1 - P(\Omega) = 0$. Again, it should be pointed out that this is not necessarily the only event with probability 0. In the die tossing experiment mentioned above, the probability of the event '6 never appears' is 0, yet that event may occur.

It is quite easy to construct a probability function when the sample space is countable. Indeed, suppose that the outcomes of the experiment are numbered $\omega_1, \omega_2, \ldots$. Assign to ω_i a non-negative weight, p_i ($i = 1, 2, \ldots$), so that

$$\sum_{i=1}^{\infty} p_i = 1 . \tag{1.3}$$

Then the probability of any event can be defined as the sum of the weights of its constituent sample points. This definition clearly satisfies axioms P1–P3. Consider again example 3: one possibility is to assign to sample point $\{i\}$ weight $1/2^i$. The events mentioned above, $A = \{1,2,3\}$, $B = \{5,6\}$, $C = \{6,7,\ldots\}$ and $D = \{4,5\}$ would then have probabilities $P(A) = 7/8$, $P(B) = 3/64$, $P(C) = 1/32$ and $P(D) = 3/32$ respectively. Note that the probabilities of A, C and D (those three events form a partition of Ω) do indeed sum up to 1.

When the sample space is uncountable (like the positive real axis in example 4), it is more difficult to give useful definitions of both events and probabilities. To treat that topic properly would involve a considerable excursion into measure theory, which is outside the scope of this book. Suffice to say that the events are the measurable subsets of Ω

and the probability function is a measure defined over those subsets. We shall assume that such a probability function is given.

If A and B are two arbitrary events, then

$$P(A \cup B) = P(A) + P(\overline{A}B) \,. \tag{1.4}$$

This is a consequence of the set identity $A \cup B = A \cup (\overline{A}B)$, plus the fact that A and $\overline{A}B$ are disjoint. Also, from $B = (AB) \cup (\overline{A}B)$ it follows that $P(B) = P(AB) + P(\overline{A}B)$. Hence,

$$P(A \cup B) = P(A) + P(B) - P(AB) \,. \tag{1.5}$$

In general, if $A_1, A_2, \ldots$ are arbitrary events, then

$$P\left[\bigcup_{i=1}^{\infty} A_i\right] \leq \sum_{i=1}^{\infty} P(A_i) \,. \tag{1.6}$$

The inequality in (1.6) becomes an equality only when $A_1, A_2, \ldots$ are disjoint.

The probability of the intersection of two events is not necessarily equal to the product of their probabilities. If, however, that happens to be true, then the two events are said to be independent of each other. Thus, A and B are independent if

$$P(AB) = P(A)P(B) \,. \tag{1.7}$$

As an illustration, take example 2, where n horses race and there are $n!$ possible outcomes. Let $\mathcal{A}$ be the set of all subsets of Ω and the probability function be generated by assigning to each of the $n!$ outcomes probability $1/n!$ (i.e. assume that all outcomes are equally likely). Suppose that $n = 3$ and consider the following events:
$A = \{(1,2,3), (1,3,2)\}$ (horse 1 wins);
$B = \{(1,2,3), (2,1,3), (2,3,1)\}$ (horse 2 finishes before horse 3);
$C = \{(1,2,3), (1,3,2), (2,1,3)\}$ (horse 1 finishes before horse 3).
Then events A and B are independent of each other, since

$$P(AB) = P(\{(1,2,3)\}) = 1/6 \,,$$

and

$$P(A)P(B) = (2/6)(3/6) = 1/6 \,.$$

However, events A and C are dependent, because

$$P(AC) = P(\{(1,2,3), (1,3,2)\}) = 2/6 \,,$$

while
$$P(A)P(C) = (2/6)(3/6) = 1/6 \ .$$

It is equally easy to verify that events B and C are dependent.

The above definition of independence reflects the intuitive idea that two events are independent of each other if the occurrence of one does not influence the likelihood of the occurrence of the other. That definition is extended recursively to arbitrary finite sets of events as follows: the n events $A_1, A_2, \ldots, A_n$ ($n > 2$) are said to be 'mutually independent', if both of the following conditions are satisfied:

(i) $P(A_1 A_2 \ldots A_n) = P(A_1) P(A_2) \ldots P(A_n)$.
(ii) Every $n - 1$ events among the $A_1, A_2, \ldots, A_n$ are mutually independent.

It should be emphasized that neither the first nor the second condition by itself is sufficient for mutual independence. In particular, it is possible that independence holds for every pair of events, yet does not hold for sets of three or more events.

1.1.3 Conditional probability

The concepts of independence and dependence are closely related to that of 'conditional probability'. If A and B are two events with positve probabilities, then the conditional probability of A, given B, is denoted by $P(A|B)$ and is defined as

$$P(A|B) = \frac{P(AB)}{P(B)} \ . \tag{1.8}$$

If A and B are independent, then $P(A|B) = P(A)$, which is consistent with the idea that the occurrence of B does not influence the probability of A.

An intuitive justification of the definition (1.8) can be given by interpreting the probability of an event as the frequency with which that event occurs when the experiment is performed a large number of times. The ratio $P(AB)/P(B)$ can then be interpreted as the frequency of occurrence of AB among those experiments in which B occurs. Hence, that ratio is the probability that A occurs, given that B has occurred.

In terms of conditional probabilities, the joint probability that A and B occur can be expressed, according to (1.8), as

$$P(AB) = P(A|B)P(B) = P(B|A)P(A) \ . \tag{1.9}$$

This formula generalizes easily to more than two events:

$$P(A_1 A_2 \ldots A_n) = \left[\prod_{i=1}^{n-1} P(A_i | A_{i+1} \ldots A_n)\right] P(A_n). \quad (1.10)$$

The probability of a given event, A, can often be determined by 'conditioning' it upon the occurrence of one of several other events. Let $B_1, B_2, \ldots$ be a complete set of events, i.e. a partition (finite or infinite) of Ω. Any event, A, can be represented as

$$A = A\Omega = A \bigcup_{i=1}^{\infty} B_i = \bigcup_{i=1}^{\infty} AB_i, \quad (1.11)$$

where the events AB_i $(i = 1, 2, \ldots)$ are disjoint. Hence,

$$P(A) = \sum_{i=1}^{\infty} P(AB_i) = \sum_{i=1}^{\infty} P(A|B_i) P(B_i). \quad (1.12)$$

This expression is known as the 'complete probability formula'. It yields the probability of an arbitrary event, A, in terms of the probabilities $P(B_i)$ and the conditional probabilities $P(A|B_i)$. We shall see numerous applications of this approach.

Alternatively, having observed that the event A has occurred, one may ask what is the probability of occurrence of some B_i. This is given by what is known as the 'Bayes formula':

$$P(B_i|A) = \frac{P(B_i A)}{P(A)} = \frac{P(A|B_i) P(B_i)}{\sum_{j=1}^{\infty} P(A|B_j) P(B_j)}. \quad (1.13)$$

Examples

5. In a group of young people consisting of 60 men and 40 women, the men divide into 20 smokers and 40 non-smokers, while the women are all non-smokers. If we know that men smokers, men non-smokers and women non-smokers survive beyond the age of 70 with probabilities 0.8, 0.85 and 0.9, respectively, what is the probability that a person chosen at random from the group will survive beyond the age of 70?

The desired quantity can be determined by applying the complete probability formula. Let A be the event 'the person chosen at random will survive beyond the age of 70'; $B1$, $B2$ and $B3$ are the events 'the person is a man smoker', 'the person is a man non-smoker' and 'the person is a woman', respectively. The three events B_1, B_2 and B_3 form a partition of Ω because one, and only one, of them occurs. Their probabilities are

$P(B_1) = 20/100 = 0.2$; $P(B_2) = 40/100 = 0.4$; $P(B_3) = 40/100 = 0.4$.
Hence we can write

$$\begin{aligned} P(A) &= P(A|B_1)P(B_1) + P(A|B_2)P(B_2) + P(A|B_3)P(B_3) \\ &= 0.8 \times 0.2 + 0.85 \times 0.4 + 0.9 \times 0.4 = 0.86 \, . \end{aligned}$$

6. In the same group of people, suppose that event A is observed, i.e. the person chosen at random survives beyond the age of 70. What is the probability that that person is a man smoker?

Now we apply Bayes' formula:

$$\begin{aligned} P(B_1|A) &= \frac{P(A|B_1)P(B_1)}{P(A|B_1)P(B_1) + P(A|B_2)P(B_2) + P(A|B_3)P(B_3)} \\ &= \frac{0.16}{0.86} \approx 0.186 \, . \end{aligned}$$

Exercises

1. Imagine an experiment consisting of tossing a coin infinitely many times. The possible outcomes are infinite sequences of 'heads' or 'tails'. Show that, with a suitable representation of outcomes, the sample space Ω is equivalent to the closed interval $[0, 1]$.

2. For the same experiment, the event 'a head appears for the first time on the ith toss of the coin' is represented by a sub-interval of $[0, 1]$. Which sub-interval?

3. An experiment consists of attempting to compile three student programs. Each program is either accepted by the compiler as valid, or is rejected. Describe the sample space Ω. Assuming that each outcome is equally likely, find the probabilities of the following events:

A: programs 1 and 2 are accepted;
B: at least one of the programs 2 and 3 is accepted;
C: at least one of the three programs is rejected;
D: program 3 is rejected.

4. For the same experiment, show that the events A and B are dependent, as are also B and C, and C and D. However, events A and D are independent. Find the conditional probabilities $P(A|B)$, $P(B|A)$, $P(C|B)$ and $P(D|C)$.

1.2 Random variables

It is often desirable to associate various numerical values with the outcomes of an experiment, whether those outcomes are themselves numeric or not. In other words it is of interest to consider functions which are defined on a sample space Ω and whose values are real numbers. Such functions are called 'random variables'. The term 'random' refers, of course, to the fact that the value of the function is not known before the experiment is performed. After that, there is a single outcome and hence a known value. The latter is called a 'realization', or an 'instance' of the random variable.

Examples

1. A life insurance company keeps the information that it has on its customers in a large database. Suppose that a customer is selected at random. An outcome of this experiment is a collection, c, of data items describing the particular customer. The following functions of c are random variables:

$X(c) =$ 'year of birth';
$Y(c) =$ '0 if single, 1 if married';
$Z(c) =$ 'sum insured';
$V(c) =$ 'yearly premium'.

2. The lifetime of a battery powering a child's toy is measured. The sample points are now positive real numbers: $\Omega = \{x : x \in \mathcal{R}^+\}$. Those points themselves can be the values of a random variable: $Y(x) = x$.

3. The execution times, x_i, of n consecutive jobs submitted to a computer are measured. This is an experiment whose outcomes are vectors, v, with n non-negative elements: $v = (x_1, x_2, \ldots, x_n)$; $x_i \geq 0$, $i = 1, 2, \ldots, n$. Among the random variables which may be of interest in this connection are:

$X(v) = \max(x_1, x_2, \ldots, x_n)$ (longest execution time);
$Y(v) = \min(x_1, x_2, \ldots, x_n)$ (shortest execution time);
$Z(v) = (x_1 + x_2 + \ldots + x_n)/n$ (sample average execution time).

4. A function which takes a fixed value, no matter what the outcome of the experiment, e.g. $X(\omega) = 5$ for all $\omega \in \Omega$, is also a random variable, despite the fact that it is not really 'random'.

* * *

Consider now an arbitrary random variable, $X(\omega)$, defined on a sample space Ω, and let (a, b) be an interval on the real line. The set, A, of sample points for which $X(\omega) \in (a, b)$ is the inverse image of (a, b) in Ω. That set represents the event 'X takes a value in the interval (a, b)'. If A is indeed an event, i.e. if $A \in \mathcal{A}$, then the probability $P(A) = P(X \in (a, b))$ is defined. We shall assume that to be the case for all intervals, finite or infinite, open or closed. It is sufficient, in fact, to require that the inverse images of all intervals of the type $(-\infty, x]$ must be events; the rest follows from axioms E1–E3.

1.2.1 Distribution functions

The probability that the random variable X takes a value which does not exceed a given number, x, is a function of x. That function, which we shall usually denote by $F(x)$, is called the 'cumulative distribution function', or just the 'distribution function', of X:

$$F(x) = P(X \leq x); \quad -\infty < x < \infty . \tag{1.14}$$

The distribution function of any random variable has the following properties.

1. If $x \leq y$ then $F(x) \leq F(y)$. This follows from the fact that the event $X \leq x$ is included in the event $X \leq y$.
2. $F(-\infty) = 0$ and $F(\infty) = 1$. This is because the event $X \leq -\infty$ is empty and the event $X \leq \infty$ is the entire sample space Ω.
3. $F(x)$ is continuous from the right, i.e. if $x_1, x_2, \ldots$ is a decreasing sequence converging to x, then $F(x) = \lim F(x_i)$. This last property is less obvious than the other two. The idea of its proof is outlined in exercise 1.

Let a and b be two real numbers, such that $a < b$. Since the events $\{X \leq a\}$ and $\{a < X \leq b\}$ are disjoint, and their union is the event $\{X \leq b\}$, we can write $P(X \leq a) + P(a < X \leq b) = P(X \leq b)$. Hence, the probability that X takes a value in the interval $(a, b]$ is given by

$$P(a < X \leq b) = F(b) - F(a) . \tag{1.15}$$

If we let $a \to b$ in this equation, we get

$$P(X = b) = F(b) - F(b^-) , \tag{1.16}$$

where $F(b^-)$ is the limit of $F(x)$ from the left, at point b.

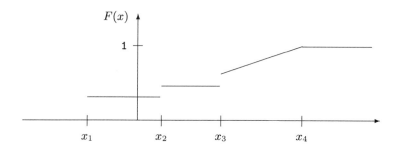

Fig. 1.1. Example of a distribution function

From equations (1.15) and (1.16) we can draw the following important conclusions:

1. If $F(a) = F(b)$, then the probability of X taking a value in the interval $(a, b]$ is 0;
2. If $F(x)$ is continuous at point b, then the probability that X takes the value b is 0 (remember that an event may have probability 0 without being impossible).

The converse of property 2 is that if a random variable takes a given value with a non-zero probability, then its distribution function has a jump at that point.

Example

5. Consider the random variable, X, whose distribution function is illustrated in figure 1.1. The shape of that function indicates that X takes values x_1, x_2 and x_3 with probabilities equal to the respective jumps at those points; the probability of taking a value in any of the open intervals $(-\infty, x_1)$, (x_1, x_2), (x_2, x_3) or (x_4, ∞) is 0; however, there is a non-zero probability that X takes a value in any sub-interval of (x_3, x_4).

* * *

A random variable which takes only a countable set of values (finite or infinite), $x_1, x_2, \ldots$, is called 'discrete'. The corresponding distribution function is constant everywhere except at points x_i, where it has jumps of magnitudes p_i, $i = 1, 2, \ldots$; those jumps satisfy the normalizing

condition
$$\sum_{i=1}^{\infty} p_i = 1 \, . \tag{1.17}$$

From now on, when we talk about the 'distribution' of a discrete random variable, that term will usually refer to the values x_i and their probabilities, p_i, rather than the function $F(x)$.

The random variables in examples 1 and 4 of this section are discrete.

A random variable whose distribution function is continuous everywhere is called 'continuous'. Such a variable can take values in any interval where its distribution function has a non-zero increment, with non-zero probability. However, the probability that it takes any particular value is zero.

The random variables in examples 2 and 3 are continuous.

The random variable in example 5 is neither discrete nor continuous, since its distribution function has both continuous portions and jumps.

1.2.2 Probability density functions

In the case of continuous random variables, the derivative of the distribution function plays an important role. That derivative is called the 'probability density function', or pdf, of the random variable. We shall usually denote it by $f(x)$. According to the definition of derivative, we can write

$$f(x) = \lim_{\Delta x \to 0} \frac{F(x + \Delta x) - F(x)}{\Delta x} = \lim_{\Delta x \to 0} \frac{P(x < X \leq x + \Delta x)}{\Delta x} \, . \tag{1.18}$$

Hence, when Δx is small, $f(x)\Delta x$ is approximately equal to the probability that the random variable takes a value in the interval $(x, x + \Delta x]$. This is what the term 'density' refers to. In what follows, we shall sometimes write that the probability that the random variable takes value x, rather than being a straightforward 0, is given by

$$P(X = x) = f(x)dx \, , \tag{1.19}$$

where dx is an infinitesimal quantity. This, of course, is nothing but a shorthand notation for equation (1.18).

A distribution function, even when it is continuous everywhere, is not necessarily differentiable everywhere. However, in all cases of interest to us, the set of points where the pdf does not exist will be finite in any finite interval.

Given a probability density function $f(x)$, the corresponding distribution function is obtained from

$$F(x) = \int_{-\infty}^{x} f(u)du . \tag{1.20}$$

In order that $F(\infty) = 1$, it is necessary that

$$\int_{-\infty}^{\infty} f(x)dx = 1 . \tag{1.21}$$

This is the continuous analogue of the normalizing condition (1.17)

Equation (1.20) allows us to express the probability that X takes a value in the interval (a, b) (for a continuous random variable it does not matter whether the end points are included or not), in terms of the probability density function:

$$P(a < X < b) = \int_{a}^{b} f(x)dx . \tag{1.22}$$

Example

6. Consider a random variable, X, which is equally likely to take any of its possible values. Such a random variable is said to be 'uniformly distributed' on its range. In the discrete case, if X can take n possible values, say $1, 2, \ldots, n$, then it is uniformly distributed on that range when

$$P(X = i) = p_i = \frac{1}{n} \; ; \; i = 1, 2, \ldots, n .$$

In the continuous case, the range of values is usually an interval on the real line: $a < X < b$. Now, 'equally likely to take any possible value' means that the probability density function of X is a constant, c, on the interval (a, b), and is 0 elsewhere:

$$f(x) = \begin{cases} c & \text{if } a < x < b \\ 0 & \text{otherwise} . \end{cases}$$

On the other hand, in order that the normalizing condition (1.21) be satisfied, the constant c must be equal to

$$c = \frac{1}{b-a} .$$

1.2 Random variables

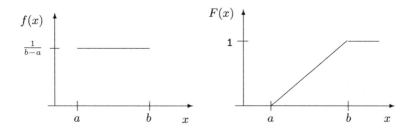

Fig. 1.2. Uniform density and distribution function

The distribution function of X is obtained from (1.20):

$$F(x) = \begin{cases} 0 & \text{if } x \leq a \\ (x-a)/(b-a) & \text{if } a < x < b \\ 1 & \text{if } b \leq x \,. \end{cases}$$

The probability density and distribution functions of a uniform random variable on an interval are shown in figure 1.2.

Of all the distribution functions on the interval (a, b), the uniform one offers least help in predicting the value that the random variable will take. If (s, t) is any sub-interval of (a, b), then the probability that X takes a value in that sub-interval depends only on its length and not on its position:

$$P(s < X < t) = F(t) - F(s) = \frac{t-s}{b-a} \,.$$

1.2.3 Joint distributions

Suppose now that X and Y are two arbitrary random variables defined on the same sample space Ω. The two-variable function $F(x, y)$, defined by

$$F(x,y) = P(X \leq x, \, Y \leq y) \,, \qquad (1.23)$$

is called the 'joint distribution function' of X and Y. Knowledge of the joint distribution function implies that of the 'marginal' distribution functions of X and Y on their own. These are denoted by $F_X(x)$ and $F_Y(y)$ respectively. We have

$$F_X(x) = P(X \leq x) = P(X \leq x, \, Y \leq \infty) = F(x, \infty) \,, \qquad (1.24)$$

(the second equality uses the fact that the intersection of the events $X \leq x$ and $Y \leq \infty$ is equal to the event $X \leq x$). Similarly,

$$F_Y(y) = P(Y \leq y) = F(\infty, y) \,. \tag{1.25}$$

In addition to characterizing X and Y separately, the joint distribution function characterizes their interaction. The two random variables are said to be 'independent of each other' if $F(x, y)$ factorizes into a product of the two marginal distribution functions, i.e. if

$$F(x, y) = F(x, \infty) F(\infty, y) = F_X(x) F_Y(y) \,. \tag{1.26}$$

In other words, in order that the random variables X and Y may be independent, the events $\{X \leq x\}$ and $\{Y \leq y\}$ must be independent, for every x and y.

When X and Y are continuous, their joint probability density function, $f(x, y)$ is defined as

$$f(x, y) = \frac{\partial^2 F(x, y)}{\partial x \partial y} \,, \tag{1.27}$$

at the points where the partial derivative exists. As in the case of a single random variable, we can interpret this by writing

$$f(x, y) = \lim_{\substack{\Delta x \to 0 \\ \Delta y \to 0}} \frac{P(x < X \leq x + \Delta x,\ y < Y \leq y + \Delta y)}{\Delta x \Delta y} \,. \tag{1.28}$$

As a shorthand for (1.28), it is possible to write

$$P(X = x,\ Y = y) = f(x, y) dx dy \,, \tag{1.29}$$

where dx and dy are infinitesimal quantities.

Given a joint pdf, $f(x, y)$, the corresponding joint distribution function is obtained from

$$F(x, y) = \int_{-\infty}^{x} \int_{-\infty}^{y} f(u, v) du dv \,. \tag{1.30}$$

In general, if S is a region in the two-dimensional (x, y) plane, the probability that the random point (X, Y) falls inside S is given by

$$P[(X, Y) \in S] = \int\int_{S} f(x, y) dx dy \,. \tag{1.31}$$

From the joint pdf, $f(x, y)$, one can derive the marginal probability density functions of X and Y:

$$f_X(x) = \int_{-\infty}^{\infty} f(x, v) dv \ ; \quad f_Y(y) = \int_{-\infty}^{\infty} f(u, y) du \,. \tag{1.32}$$

1.2 Random variables

The first of these equations is obtained by setting $y = \infty$ in (1.30) and then differentiating with respect to x; the second by setting $x = \infty$ and differentiating with respect to y.

It is also of interest to consider the 'conditional density function' of X given Y. This is defined as

$$f_{X|Y}(x|y) = \begin{cases} f(x,y)/f_Y(y) & \text{if } f_Y(y) > 0 \\ 0 & \text{if } f_Y(y) = 0 \end{cases} \tag{1.33}$$

Taken as a function of x, for a fixed y, the right-hand side of (1.33) is a probability density function for the random variable X, given that the random variable Y has taken value y. This can be justified by using the conditional probability formula (1.8), together with expressions (1.19) and (1.29):

$$P(X = x \mid Y = y) = \frac{P(X = x, Y = y)}{P(Y = y)} = \frac{f(x,y)dxdy}{f_Y(y)dy} = \frac{f(x,y)}{f_Y(y)}dx \ .$$

The condition for independence (1.26) can be stated in terms of probability density functions: two (continuous) random variables X and Y are mutually independent if

$$f(x,y) = f_X(x)f_Y(y) \ . \tag{1.34}$$

As can be expected, if X and Y are independent, then the conditional pdf of X given Y is the same as the unconditional pdf of X:

$$f_{X|Y}(x|y) = f_X(x) \ . \tag{1.35}$$

Example

7. Consider an experiment consisting of tossing two true dice. All 36 possible outcomes are equally likely. Let X, Y and Z be the random variables representing the numbers shown on the first die, the second die and the total on both dice, respectively. Then, since $P(X = i) = 1/6$, $P(Y = j) = 1/6$ and $P(X = i, Y = j) = 1/36$ $(i, j = 1, \ldots, 6)$, it is not difficult to see that the joint distribution function of X and Y factorizes into a product of the marginal distribution functions. Hence, X and Y are independent random variables. However, X and Z are dependent, because $P(X \leq 1, Z \leq 2) = 1/36$, whereas $P(X \leq 1)P(Z \leq 2) = 1/216$.

1.2.4 Largest and smallest

Suppose that $X_1, X_2, \ldots, X_n$ are independent random variables with distribution functions $F_1(x), F_2(x), \ldots, F_n(x)$, respectively. It is often of interest to consider the largest and smallest values that occur among them:

$$X_{max} = \max(X_1, X_2, \ldots, X_n) \; ; \; X_{min} = \min(X_1, X_2, \ldots, X_n) . \tag{1.36}$$

For example, $X_1, X_2, \ldots, X_n$ may be the execution times of n programs which start running in parallel, or the lifetimes of n machines which start working in parallel (and are not replaced when broken). Then X_{max} is the time it takes for all programs to complete, or all machines to fail; X_{min} is the time until the first completion, or the first failure.

Denote the distribution functions of X_{max} and X_{min} by $F_{max}(x)$ and $F_{min}(x)$, respectively. To find $F_{max}(x)$, note that in order for X_{max} to be less than or equal to x, all random variables must be less than or equal to x. Hence,

$$F_{max}(x) = \prod_{i=1}^{n} F_i(x) . \tag{1.37}$$

In order for X_{min} to be less than or equal to x, it *must not* be true that all random variables are greater than or equal to x. Therefore, $F_{min}(x)$ is given by

$$F_{min}(x) = 1 - \prod_{i=1}^{n} [1 - F_i(x)] . \tag{1.38}$$

More generally, let Y_i be the ith smallest value observed among the X_i (e.g. $Y_1 = X_{min}$, $Y_n = X_{max}$). The random variables $Y_1, Y_2, \ldots, Y_n$ are called the 'order statistics' of the sample $X_1, X_2, \ldots, X_n$. The distribution of Y_i is difficult to obtain in the general case, but reasonably simple expressions can be derived when the random variables X_i are identically distributed, i.e. when $F_i(x) = F(x)$ $(i = 1, 2, \ldots, n)$. This is done in example 5 of section 1.4.2.

Exercises

1. Let X be a random variable with probability distribution function $F(x)$, and let $x_1, x_2, \ldots$ be a monotone decreasing sequence converging

1.2 Random variables

to x. For $i = 1, 2, \ldots$, let A_i be the events defined by $A_1 = \{X > x_1\}$; $A_i = \{x_i < X \leq x_{i-1}\} (i > 1)$. Show that

$$P(X > x_n) = \sum_{i=1}^{n} P(A_i),$$

and

$$P(X > x) = \sum_{i=1}^{\infty} P(A_i).$$

Hence prove that $F(x) = \lim_{n \to \infty} F(x_n)$.

2. The distribution function of the 'battery lifetime' random variable, Y, from example 2, is given by

$$F(x) = \begin{cases} 0 & \text{if } x < 0 \\ 0.01x^2 & \text{if } 0 \leq x \leq 10 \\ 1 & \text{if } x > 10 . \end{cases}$$

Find the probability that the battery lifetime is either less than 1 or greater than 9. If an 'adjusted lifetime' random variable, Z, is defined by

$$Z = \begin{cases} Y & \text{if } Y \leq 6 \\ Y - 1 & \text{if } X > 6, \end{cases}$$

find the distribution function of Z. Is Z continuous?

3. Let X and Y be two random variables with joint probability distribution function $F(x, y)$, and let S be the rectangle $S = \{(x, y) : a < x \leq b, c < y \leq d\}$. Show that the probability that the random point (X, Y) falls inside S is given by

$$P[(X, Y) \in S] = F(b, d) - F(a, d) - F(b, c) + F(a, c) .$$

(Hint: find first the probabilities $P(a < X \leq b, Y \leq d)$ and $P(a < X \leq b, Y \leq c)$.)

4. Let X and Y be two random variables with joint pdf $f(x, y)$. Using the definition of conditional probability density function, obtain an expression for $f_{Y|X}(y|x)$ in terms of $f_Y(y)$ and $f_{X|Y}(x|y)$ only (that expression is the continuous analogue of the Bayes formula (1.13)).

5. The random variables $X_1, X_2, \ldots, X_{10}$ are independent and identically uniformly distributed on the interval $(0, 1)$. Find the probabilities that the smallest of them is less than 0.1, and the largest is greater than 0.8, respectively.

1.3 Expectation and other moments

A random variable is characterized completely by its probability distribution function (or by its probability density function). However, such a full characterization is not always easy to obtain, nor is it always absolutely necessary. It is sometimes desirable, and sufficient for practical purposes, to describe a random variable by one, or several, numbers which somehow summarize its essential attributes. Descriptors of this type exist. They are defined in terms of the distribution function, but can often be determined directly, without having to know that function.

One of the most important characteristics of a random variable, X, is its 'mean', or 'average', or 'expectation', denoted by $E(X)$. This is formally defined as the integral

$$E(X) = \int_{-\infty}^{\infty} x\, dF(x) , \qquad (1.39)$$

where $F(x)$ is the probability distribution function of X.

One can give a physical interpretation to the above quantity, and at the same time provide an intuitive justification for calling it a 'mean'. Let us replace, temporarily, the word 'probability' with the word 'mass', and think of $F(x)$ as describing the distribution of one unit of mass along the real line. That is, the mass is spread in such a way that $F(x)$ of it is to the left of, or at, point x, for every real x. Then the centre of mass for the resulting line is precisely at the point given by the right-hand side of (1.39).

If X is a discrete random variable, taking values $x_1, x_2, \ldots$ with probabilities $p_1, p_2, \ldots$ respectively (in the physical analogy, the mass is concentrated at points $x_1, x_2, \ldots$, in amounts $p_1, p_2, \ldots$), then the integral in (1.39) becomes a (finite or infinite) sum:

$$E(X) = \sum_{i=1}^{\infty} x_i p_i . \qquad (1.40)$$

This expression is easier to see as a mean value: a weighted average of all the values that the random variable can take, where the weight of each value is the probability with which it is taken.

If X is a continuous random variable with pdf $f(x)$, then (1.39) can

1.3 Expectation and other moments

be rewritten as

$$E(X) = \int_{-\infty}^{\infty} xf(x)dx . \tag{1.41}$$

This is the continuous analogue of (1.40), since, according to (1.19), X takes value x with probability $f(x)dx$.

If X is neither discrete nor continuous, i.e. if $F(x)$ has continuously changing portions as well as jumps, then the general integral (1.39) has to be employed. Its evaluation in most practical cases reduces to that of terms like

$$\int_{a_i}^{b_i} xf(x)dx ,$$

over the continuous portions, plus terms of type

$$x_i[F(x_i) - F(x_i^-)] ,$$

at points where $F(x)$ has jumps.

Implicit in the definition of expectation is an assumption that the integral in the right-hand side of (1.39) converges. If it does not, then we say that $E(X)$ does not exist.

Examples

1. A gambler pays £3 for the privilege of throwing a single die. If the number that comes up is greater than 3, he or she will win that number of pounds; otherwise will get nothing. Let X be the difference between the money gained and spent. This is a random variable whose value is -3 with probability $1/2$ (if 1, 2 or 3 comes up), 1 with probability $1/6$ (if 4 comes up), 2 with probability $1/6$ (if 5 comes up) and 3 with probability $1/6$ (if 6 comes up). We are assuming that the six possible outcomes are equally likely. Hence, the expected value of X is

$$E(X) = -\frac{3}{2} + \frac{1}{6} + \frac{2}{6} + \frac{3}{6} = -\frac{1}{2} ,$$

i.e. the gambler will lose 50 pence on the average.

2. A random variable, X, takes value i with probability $1/2^i$ ($i = 1, 2, \ldots$). For instance, X could be the number of task execution attempts in example 3 of section 1.1. The expectation of X is obtained as

$$E(X) = \sum_{i=1}^{\infty} \frac{i}{2^i} = 2 .$$

3. A random variable, X, is distributed uniformly on the interval (a, b) (see example 6 in section 1.2). Its probability density function is, therefore,

$$f(x) = \begin{cases} 1/(b-a) & \text{if } a < x < b \\ 0 & \text{otherwise} \end{cases}.$$

The average value of X is

$$E(X) = \int_a^b \frac{x}{b-a} dx = \frac{1}{b-a}\left[\frac{b^2}{2} - \frac{a^2}{2}\right] = \frac{a+b}{2}.$$

4. Let X be a random variable whose pdf is

$$f(x) = \frac{1}{\pi(1+x^2)} \; ; \; -\infty < x < \infty.$$

This is known as the 'Cauchy density'. The variable X can be interpreted as the tangent of an angle (in radians) drawn at random from the range $(-\pi/2, \pi/2)$. This random variable has no expectation, since the integral

$$\frac{1}{\pi} \int_{-\infty}^{\infty} \frac{xdx}{1+x^2},$$

does not converge.

* * *

The notions of probability and expectation are, in a certain sense, equally fundamental. To show that, consider an arbitrary event, A, in the sample space Ω. Define a random variable $I_A(\omega)$, for $\omega \in \Omega$, as follows:

$$I_A(\omega) = \begin{cases} 1 & \text{if } \omega \in A \\ 0 & \text{otherwise} \end{cases}. \tag{1.42}$$

In other words, I_A takes value 1 if A occurs and 0 if it does not. That random variable is called the 'indicator' of the event A. An application of (1.40) shows immediately that the expectation of I_A is equal to the probability of A:

$$E(I_A) = 1 \times P(I_A = 1) + 0 \times P(I_A = 0) = P(I_A = 1) = P(A). \tag{1.43}$$

Thus, instead of starting with an axiomatic treatment of probability and then defining expectation as in (1.39), one could take expectation as a starting point and define probability by means of (1.43).

1.3.1 Properties of expectation

The following properties of expectation are simple consequences of the definition (1.39). They would be the defining axioms if expectation were to be taken as the fundamental concept.

1. If $X \geq 0$ then $E(X) \geq 0$.
2. If c is a constant, then $E(cX) = cE(X)$.
3. $E(X + Y) = E(X) + E(Y)$, for any random variables X and Y.
4. $E(1) = 1$.
5. If $X_1, X_2, \ldots$ is a monotonic sequence of random variables, converging in distribution to X, then $E(X) = \lim_{n \to \infty} E(X_n)$.

Property 1 is obvious: if X takes only non-negative values, then the integral in (1.39) can be taken over the semi-axis $0 \leq x < \infty$ and is therefore non-negative.

Property 2 follows from the observation that the random variable cX takes the value cx with the same probability as the one with which X takes the value x. Hence,

$$E(cX) = \int_{-\infty}^{\infty} cx dF(x) = cE(X) .$$

We shall demonstrate property 3 for the case when X and Y are continuous, with joint pdf $f(x, y)$. The general case is treated similarly. Bearing in mind that when X takes the value x and Y takes the value y (which happens with probability $f(x,y)dxdy$), $X + Y$ takes the value $x + y$, we can write

$$\begin{aligned} E(X+Y) &= \int_{-\infty}^{\infty}\int_{-\infty}^{\infty} (x+y)f(x,y)dxdy \\ &= \int_{-\infty}^{\infty} xf_X(x)dx + \int_{-\infty}^{\infty} yf_Y(y)dy = E(X) + E(Y) , \end{aligned}$$

where the second equality relies on (1.32). Note that property 3 holds regardless of whether X and Y are independent.

Property 4 follows immediately from (1.40), which now has a single term.

Property 5 is a consequence of the fact that limit and integration can be interchanged.

From property 1 it follows that inequalities between random variables are preserved after taking expectations. In other words, if $X \leq Y$, then $E(X) \leq E(Y)$. However, the reverse implication does not necessarily hold.

Properties 2 and 3 imply that if $X_1, X_2, \ldots, X_n$ are random variables (dependent or independent), and $c_1, c_2, \ldots, c_n$ are constants, then

$$E\left[\sum_{i=1}^{n} c_i X_i\right] = \sum_{i=1}^{n} c_i E(X_i) \, .$$

From the definition (1.39) it follows that if X and Y are independent random variables, i.e. if $F_{X,Y}(x,y) = F_X(x)F_Y(y)$, then

$$E(XY) = E(X)E(Y) \, . \tag{1.44}$$

However, the validity of (1.44) does not imply that X and Y are independent of each other.

There are other ways of attempting to capture the essence of 'central value' of a random variable X. For example, one can define the 'median', m, as the point such that the events $X < m$ and $X > m$ are equally likely. In other words, m is defined as the solution of the equation $F(m) = 1/2$ (and if there is a whole interval satisfying that equation, m is its mid-point). Or one can introduce the 'mode' of X, as the value most likely to be taken (in the continuous case, the value with the largest probability density). However, the expectation is alone among these characteristics in having the above properties of a linear operator.

The expectation is also known as the 'first moment' of a random variable. More generally, the nth moment of X, M_n, is defined as

$$M_n = E(X^n) = \int_{-\infty}^{\infty} x^n dF(x) \, . \tag{1.45}$$

In the discrete and continuous special cases, M_n is evaluated by expressions similar to (1.40) and (1.41), respectively.

It can be shown that, under certain quite general conditions, a random variable is completely characterized by the set of all its moments. In other words, if M_n is known for all $n = 1, 2, \ldots$, then the distribution function of X is determined uniquely.

1.3.2 Variance and covariance

The second moment, M_2, and quantities related to it, are particularly important. Thus, a fundamental characteristic of a random variable, X, is its 'variance', denoted by $\mathrm{Var}(X)$. This is a measure of the 'spread', or 'dispersion', of the random variable around its mean. More precisely,

1.3 Expectation and other moments

the variance is defined as the average squared distance between X and its mean:

$$\text{Var}(X) = E([X - E(X)]^2) = E(X^2) - [E(X)]^2 = M_2 - M_1^2 . \quad (1.46)$$

From this definition, and from the fact that the variance is always non-negative, it follows that $M_2 \geq M_1^2$, for any random variable X. Equality is reached when X is a constant, c. Then we have

$$[E(c)]^2 = c^2 = E(c^2) \; ; \; \text{Var}(c) = 0 . \quad (1.47)$$

Clearly, if X takes values close to the mean $E(X)$ with high probability, then the variance $\text{Var}(X)$ is small. Much less obvious, but also true, is the converse assertion: if $\text{Var}(X)$ is small, then X takes values close to $E(X)$ with a high probability. This statement is quantified by the following result, known as 'Chebyshev's inequality':

$$P(|X - E(X)| < d) \geq 1 - \frac{\text{Var}(X)}{d^2} . \quad (1.48)$$

To prove (1.48), consider the event $A = \{|X - E(X)| < d\}$, and let I_A be its indicator. Then, whatever the outcome of the experiment,

$$I_A \geq 1 - \frac{[X - E(X)]^2}{d^2} .$$

(If A occurs, then I_A is 1 and the right-hand side is less than 1; if A does not occur, then I_A is 0 and the right-hand side is non-positive). Taking expectations, and remembering that $E(I_A) = P(A)$, yields (1.48).

A trivial corollary of Chebyshev's inequality is that, if $\text{Var}(X)$ is 0, then X is equal to its mean with probability 1, i.e. X is almost certainly a constant.

The variance has the following properties, which follow directly from the definition (1.46):

1. If c is a constant, then $\text{Var}(cX) = c^2 \text{Var}(X)$.
2. If X and Y are independent random variables, then $\text{Var}(X+Y) = \text{Var}(X) + \text{Var}(Y)$.

Since any constant, c, is independent of any random variable, X, property 2 implies that

$$\text{Var}(X + c) = \text{Var}(X) . \quad (1.49)$$

The ratio of the variance to the square of the mean is also sometimes used as a measure of dispersion. That ratio is called the 'squared coefficient of variation', and is denoted by C^2:

$$C^2 = \frac{\text{Var}(X)}{[E(X)]^2} = \frac{M_2}{M_1^2} - 1. \tag{1.50}$$

The square root of the variance, $\sigma = \sqrt{\text{Var}(X)}$, is called the 'standard deviation' of the random variable X.

The 'covariance' between two random variables, X and Y, is denoted by $\text{Cov}(X,Y)$, and is defined as

$$\begin{aligned}\text{Cov}(X,Y) &= E\{[X - E(X)][Y - E(Y)]\} \\ &= E(XY) - E(X)E(Y).\end{aligned} \tag{1.51}$$

This is a measure of the dependency, or rather the 'correlation', between X and Y. According to (1.44), if the two random variables are independent, their covariance is 0. However, the converse is not necessarily true.

The covariance is sometimes normalized, to produce a value within the range $[-1,1]$. The resulting quantity is called the 'correlation coefficient', $r(X,Y)$, and is defined as

$$r(X,Y) = \frac{\text{Cov}(X,Y)}{\sqrt{\text{Var}(X)\text{Var}(Y)}}. \tag{1.52}$$

The random variables X and Y are said to be 'positively correlated', 'negatively correlated', or 'uncorrelated', if $r(X,Y) > 0$, $r(X,Y) < 0$, or $r(X,Y) = 0$, respectively. If there is a strict linear relationship between them, i.e. if there are constants a and b such that $X = aY + b$, then $r(X,Y) = 1$ if $a > 0$, and $r(X,Y) = -1$ if $a < 0$. This follows directly from (1.51) and the properties of variance. In general, $-1 \leq r(X,Y) \leq 1$ (see exercise 4).

Using (1.46) and (1.51), we can write an expression for the variance of the sum of two arbitrary random variables, X and Y:

$$\text{Var}(X+Y) = \text{Var}(X) + \text{Var}(Y) + 2\text{Cov}(X,Y). \tag{1.53}$$

This expression shows that, in order for the additive property of variance to hold (property 2 above), it is enough that X and Y should be uncorrelated; there is no need to require independence.

Exercises

1. Find the second moment and the variance of the uniformly distributed random variable defined in example 6 of section 1.2.

1.3 Expectation and other moments

2. Let X be a non-negative random variable and d a positive constant. Consider the event $A = \{X > d\}$ and show that its indicator satisfies the inequality $I_A \leq X/d$. Hence derive the 'Markov inequality'

$$P(X > d) \leq \frac{E(X)}{d}.$$

3. Show that, for any two random variables X and Y,

$$[E(XY)]^2 \leq E(X^2)E(Y^2).$$

This is known as the Cauchy–Schwarz inequality. (Hint: note that

$$E(X^2) - 2aE(XY) + a^2 E(Y^2) = E[(X - aY)^2]$$

is non-negative for every a. Hence, the discriminant of the quadratic form in the left-hand side must be less than or equal to 0.)

4. Show that the correlation coefficient of X and Y, $r(X,Y)$, is always in the range $-1 \leq r(X,Y) \leq 1$. (Hint: apply the Cauchy–Schwarz inequality established in the previous exercise to the random variables $X - E(X)$ and $Y - E(Y)$.)

5. Let X and Y be two random variables and a and b be two constants. Show that

$$\mathrm{Var}(aX + bY) = a^2 \mathrm{Var}(X) + 2ab\mathrm{Cov}(X,Y) + b^2 \mathrm{Var}(Y).$$

Generalize this expression to a linear combination of n random variables.

6. Show that if the expectation $E(X)$ exists, it can be obtained from

$$E(X) = \int_0^\infty [1 - F(x)]dx - \int_{-\infty}^0 F(x)dx.$$

(Hint: integrate the above expression by parts and reduce it to (1.39).)

7. Show that if the second moment, M_2, exists, it can be obtained from

$$M_2 = 2\int_0^\infty x[1 - F(x)]dx - 2\int_{-\infty}^0 xF(x)dx.$$

1.4 Bernoulli trials and related random variables

We shall examine here what is perhaps the simplest non-trivial example of a random process, i.e. an experiment which involves time and where different things may happen at different moments. Imagine first an action which may have two possible outcomes, such as tossing a coin, or exposing the top card of a well-shuffled pack in order to see whether its rank is higher than 10 or not. We shall call such an action a 'trial' and its two outcomes 'success' and 'failure'.

Now consider an experiment which consists of performing a sequence of identical trials, so that the outcome of trial i is independent of the outcomes of trials $1, 2, \ldots, i-1$. Moreover, suppose that the outcome of each trial is a success with probability q and a failure with probability $1-q$ ($0 < q < 1$). Such an experiment is called 'Bernoulli trials'; its outcomes are sequences (finite or infinite) of successes and failures.

1.4.1 The geometric distribution

A random variable of interest in a Bernoulli trials experiment is the index, K, of the trial at which the first success occurs. Because of the nature of the experiment, the number of trials between the first and the second successes (excluding the former but including the latter), or between any two consecutive successes, has the same distribution as K.

To find the probability, p_k, that K takes the value k, note that the event $\{K = k\}$ occurs if, and only if, the first $k-1$ trials are failures and the kth one is a success. Hence,

$$p_k = P(K = k) = (1-q)^{k-1} q \; ; \quad k = 1, 2, \ldots . \quad (1.54)$$

From this expression, the distribution function of K is obtained as

$$F(k) = P(K \leq k) = \sum_{i=1}^{k} p_i = 1 - (1-q)^k \; ; \quad k = 1, 2, \ldots . \quad (1.55)$$

This is known as the 'geometric distribution'. That it is, indeed, a distribution, follows from the fact that

$$\lim_{k \to \infty} F(k) = \sum_{i=1}^{\infty} p_k = 1 .$$

In other words, a success will occur eventually with probability 1. The experiment may, of course, result in an infinite sequence consisting entirely of failures. However, the probability of such an outcome is 0.

1.4 Bernoulli trials and related random variables

The average value of K is given by

$$E(K) = \sum_{k=1}^{\infty} k p_k = q \sum_{k=1}^{\infty} k(1-q)^{k-1} = \frac{1}{q}. \tag{1.56}$$

As can be expected, the lower the probability of success, the longer one has to wait, on the average, until the first success.

The second moment and the variance of the random variable K are obtained in a straightforward manner from their definitions. They are given by

$$E(K^2) = \frac{2-q}{q^2} \; ; \; \text{Var}(K) = \frac{1-q}{q^2}. \tag{1.57}$$

The geometric distribution is unique among the discrete distributions in that it has the so-called 'memoryless property': the fact that no successes have occurred during a set of i consecutive trials does not affect the likelihood of no successes occurring during a subsequent set of j consecutive trials. Formally, this can be stated as follows:

$$P(K > i+j \mid K > i) = \frac{1 - F(i+j)}{1 - F(i)} = (1-q)^j = P(K > j). \tag{1.58}$$

Thus, the events 'no successes occur during i consecutive trials' and 'no successes occur during the next j consecutive trials' are independent of each other. Conversely, if (1.58) holds for some discrete distribution, $F(k)$, $k = 1, 2, \ldots$, then that distribution must have the form (1.54), i.e. it must be geometric (see exercise 1).

Examples

1. A processor has an infinite supply of jobs, which it executes one after the other. The execution of a job consists of giving it one or more quanta of service, each lasting exactly one time unit. The job is completed at the end of a service quantum with probability q, and requires more service with probability $1 - q$ ($0 < q < 1$); each such outcome is independent of all previous ones (figure 1.3).

The consecutive service quanta can thus be considered as Bernoulli trials, a job completion being a success. Therefore, the number of service quanta required by a job, and hence the job execution time, is distributed geometrically with parameter q. According to the memoryless property, the probability that a job will require at least j more time units to complete, given that it has already been running for i time units, is the same as the unconditional probability that it needs at least j time units.

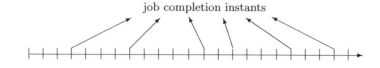

Fig. 1.3. Jobs receiving service in quanta

Fig. 1.4. A slotted communication channel

The average execution time of a job, as well as the average *remaining* execution time given that it has already received some service, is equal to $1/q$ quanta.

2. Consider a 'slotted' communication channel which can accept packets of information for transmission at time instants $0, 1, 2, \ldots$. A packet takes exactly one time unit to transmit. At time i, there is either one packet available for transmission, with probability q, or no packets, with probability $1-q$, regardless of what happened at previous instants (figure 1.4).

The channel goes through alternating periods of being idle (having nothing to transmit), and busy (transmitting a sequence of packets). Suppose that the first slot is an idle one, i.e. there is no packet at time 0. Then, considering the instants $1, 2, \ldots$ as Bernoulli trials, success occurring when a packet is present, we see that the length of the first idle period is determined by the index of the first success. That length, and hence the length of any other idle period, is therefore distributed geometrically with parameter q. Similarly, treating the absence of a packet as a success, we conclude that the duration of any busy period is distributed geometrically with parameter $1 - q$.

3. In the above example, let us change the definition of busy and idle periods by postulating that every busy slot is a separate busy period (of length 1). Between two consecutive busy periods, there is an idle period which may be of length 0. Now, if J is the length of an idle period, J takes value j when j idle slots are followed by a busy slot ($j = 0, 1, \ldots$).

1.4 Bernoulli trials and related random variables

Hence,
$$P(J = j) = (1-q)^j q \;\; ; \;\; j = 0, 1, \ldots,$$
and
$$P(J \le j) = 1 - (1-q)^{j+1} \;\; ; \;\; j = 0, 1, \ldots.$$

We shall sometimes refer to this as the 'modified geometric' distribution. Clearly, the geometric random variable, K, and the modified geometric one, J, are closely related. Indeed, a simple comparison shows that the random variable $K - 1$ has the same distribution as J. Hence, the mean and variance of J are given by:

$$E(J) = E(K) - 1 = \frac{1-q}{q} \;\; ; \;\; \mathrm{Var}(J) = \mathrm{Var}(K) = \frac{1-q}{q^2}.$$

1.4.2 The binomial distribution

Another random variable of interest in connection with a Bernoulli trials experiment is the number of successes, S, that occur during the first n trials (or during any n trials). If we select a particular set of j trials ($j \le n$), then the probability that those trials are successes and the other $n - j$ trials are failures is equal to $q^j(1-q)^{n-j}$. On the other hand, the number of ways of selecting j out of n possibilities is equal to the binomial coefficient

$$\binom{n}{j} = \frac{n!}{j!(n-j)!}.$$

Hence, the probability, $p_{j,n}$, that there are j successes among n trials is given by

$$p_{j,n} = \binom{n}{j} q^j (1-q)^{n-j} \;\; ; \;\; j = 0, 1, \ldots, n. \tag{1.59}$$

This is known as the 'binomial distribution'.

Examples

4. Multiple modular redundancy. In a system where it is important to minimize the effect of breakdowns, every task is replicated on n identical and independent processors. If more than half of those processors produce the same result, the task is said to have completed successfully. Assume that each processor is correct on a given task with probability q, and fails with probability $1 - q$. Suppose, for simplicity, that n is an odd number, $n = 2k+1$, so that $k+1$ correct processors are enough for a majority. Determine the following:

(a) The probability, A, that a given task completes successfully.
(b) The conditional probability, B_j, that the task completes successfully, given that at most j processors are working correctly ($j > k$).
(c) The conditional probability, C_j, that at least j processors are working correctly, given that a task completes successfully.

Let S be the number of correct processors. This random variable has the binomial distribution (1.59). We can therefore write

$$A = P(S > k) = \sum_{s=k+1}^{n} p_{s,n} \ .$$

$$B_j = P(S > k \mid S \le j) = \frac{\sum_{i=k+1}^{j} p_{i,n}}{\sum_{i=0}^{j} p_{i,n}} \ .$$

$$C_j = P(S \ge j \mid S > k) = \frac{\sum_{i=\max(j,k+1)}^{n} p_{i,n}}{\sum_{i=k+1}^{n} p_{i,n}} \ .$$

5. Order statistics. Let $X_1, X_2, \ldots, X_n$ be independent random variables, all having the same distribution function, $F(x)$. Denote by Y_i the ith smallest value observed among these variables ($i = 1, 2, \ldots, n$; see section 1.2.4). To determine the corresponding distribution function, $G_i(x)$, we argue as follows. In order for Y_i to be less than or equal to x, at least i of the X-variables must be less than or equal to x. Treating each variable as a Bernoulli trial where a value less than or equal to x is a 'success', we conclude that the number of variables that are less than or equal to x has the binomial distribution with parameter $F(x)$. Therefore,

$$G_i(x) = \sum_{j=i}^{n} \binom{n}{j} [F(x)]^j [1 - F(x)]^{n-j} \ . \tag{1.60}$$

If the variables X_i have a pdf, $f(x)$, then so do the order statistics, Y_i, for all i; denote the latter pdfs by $g_i(x)$. Those densities can be found either by differentiating (1.60), or by noting that, in order for Y_i to take value x, one of the X-variables must take value x, $i-1$ other variables

1.4 Bernoulli trials and related random variables

must be less than or equal to x, and the rest must be greater than x. This leads to

$$g_i(x) = n \binom{n-1}{i-1} f(x)[F(x)]^{i-1}[1-F(x)]^{n-i} . \tag{1.61}$$

* * *

The mean and variance of the binomial distribution can be obtained directly from (1.59). However, a slightly circuitous route turns out to be easier and more elegant. The first step is to introduce the random variable X_i, representing the number of successes that occur on the ith trial. Obviously, X_i can take the values 1 and 0 only, with probabilities q and $1-q$ respectively, independently of i. The mean, second moment and variance of X_i are easily derived:

$$E(X_i) = E(X_i^2) = P(X_i = 1) = q \; ; \; \text{Var}(X_i) = q(1-q) . \tag{1.62}$$

Now, the number of successes among n trials can be expressed as

$$S = \sum_{i=1}^{n} X_i . \tag{1.63}$$

Taking expectations in (1.63) we obtain

$$E(S) = nq . \tag{1.64}$$

Similarly, since the random variables X_i are mutually independent, the variance of their sum is equal to

$$\text{Var}(S) = nq(1-q) . \tag{1.65}$$

The binomial distribution plays a very important role in both probability theory and modelling. Consider the ratio S/n, which represents the frequency with which successes occur within the sequence of n trials. The expected value of that ratio is, according to (1.64), equal to q for any given n. This by itself does not, of course, imply anything about the value that S/n will take in a particular realization of the experiment. However, it is an empirical fact that, if n is large, the random variable S/n is likely to take a value which is close to the probability of success, q. Indeed, before the axiomatic treatment of probability was developed, the natural scientists used to define the probability of an event as the long-term frequency with which the event occurred. We can now supply a mathematical confirmation of this empirical fact. It can be shown that, whatever the positive number ε, the probability that S/n deviates from

q by less than ε approaches 1 as n increases:

$$\lim_{n\to\infty} P\left(\left|\frac{S}{n} - q\right| < \varepsilon\right) = 1 \, . \tag{1.66}$$

The proof of this assertion, known as the 'weak law of large numbers', is outlined in exercise 3. Note that, in view of (1.63), S/n is the 'arithmetic mean', or the 'sample mean', of the random variables $X_1, X_2, \ldots, X_n$. If those were arbitrary independent random variables (rather than being indicators of success), with mean q and finite variance, (1.66) would continue to hold. In fact, an even stronger statement can be made. We can 'almost' guarantee that the sample mean S/n will converge to the probability q in any given realization of the experiment:

$$P\left(\lim_{n\to\infty} \frac{S}{n} = q\right) = 1 \, . \tag{1.67}$$

In other words, although there may be experimental outcomes for which S/n does not converge to q, the probability of such outcomes is 0. This is known as the 'strong law of large numbers'.

In example 1 above, the number of jobs completed by the server during n service quanta has the binomial distribution. The expectation of that number is nq and hence the average system throughput is q jobs per unit time. Similarly, in example 2, the number of packets transmitted during n slots is binomially distributed and the average channel throughput is q packets per unit time.

There are many applications of the binomial distribution in the areas of quality control and reliability. If each one of a series of items has a probability q of being defective, independently of the others, and a sample of n items is given, then the number, S, of defective items in that sample is binomially distributed. Moreover, according to the laws of large numbers, the fraction S/n can be taken as an estimate of q. In this context, 'item' may mean a manufactured object, a hardware component or a software module.

It is sometimes possible to generalize the Bernoulli trials scheme by allowing the probability of success to depend on the number of successes that have occurred. Results associated with intervals between successes can still be obtained. Consider, for example, the following simple model.

Example

6. Software reliability growth. A large software product is repeatedly tested, by running it with different inputs. A test fails if it reveals

1.4 Bernoulli trials and related random variables

a fault; otherwise it is successful. After a failed test, the exposed fault is identified and fixed; then the testing continues. Suppose that the initial probability of failure is equal to q_0. That probability remains unchanged so long as the tests are successful. When the first fault is revealed and removed, the probability of failure becomes q_1. In general, after the removal of the i th fault, the probability of a test failure is q_i ($i = 0, 1, \ldots$). Thus, the number of tests that are performed beween the discovery of the i th and $i + 1$ st faults is distributed geometrically with parameter $1 - q_i$.

If the act of removing a fault does not introduce new ones, the reliability of the software should grow, i.e. the probabilities q_i should decrease with i. Different reliability growth models make different assumptions concerning those probabilities. For instance, they may be assumed to decrease linearly: $q_i = (m - i)/n$ for $i = 0, 1, \ldots, m$ and $q_i = 0$ for $i > m$, where m and n are some positive integers with $n > m$. Alternatively, the failure probabilities may decrease exponentially: $q_i = q_0 \alpha^i$ ($i = 0, 1, \ldots, 0 < \alpha < 1$).

A random variable of interest in this context is the number of tests, N_j, that have to be carried out until the discovery of the j th fault. The mean and variance of N_j can be obtained by remarking that that number is equal to the sum of j independent geometrically distributed random variables, the i th of which has parameter $1 - q_i$:

$$E(N_j) = \sum_{i=0}^{j-1} \frac{1}{1 - q_i} \; ; \; \text{Var}(N_j) = \sum_{i=0}^{j-1} \frac{q_i}{(1 - q_i)^2} .$$

In the special case where all q_i are equal, the distribution of N_j is also easily derived (see exercise 4).

Exercises

1. Let $a_1, a_2, \ldots$ be a sequence of numbers in the interval $(0, 1)$, which satisfy the equations $a_{i+j} = a_i a_j$, for $i, j = 1, 2, \ldots$. Show, by setting $j = 1$ and solving the resulting recurrence equations, that $a_i = a_1^i$, $i = 1, 2, \ldots$. Hence demonstrate that any distribution which satisfies (1.58) must be of the form (1.54).

2. Establish (1.66) by applying Chebyshev's inequality to the random variable S/n. (Hint: the variance of S/n approaches 0 as n increases.)

3. In order to assess how well a student has mastered a new programming language, she is asked to write and run five unrelated programs. She will be considered to have passed the test if at least three of them are correct. Assuming that each program the student writes is correct with probability 0.6 (independently of the others), what is the probability that she will pass?

4. In a Bernoulli trials experiment with a probability of success q, let N_j be the index of the trial at which the jth success occurs, i.e. the number of trials required to achieve j successes. Bearing in mind that, in order for N_j to take value n, the nth trial must be a success and there must be exactly $j-1$ successes among the preceding $n-1$ trials, show that the distribution of N_j, known as the 'negative binomial distribution', is given by

$$P(N_j = n) = \binom{n-1}{j-1} q^j (1-q)^{n-j} \; ; \; n = j, j+1, \ldots \; .$$

Hence (or using the fact that N_j is the sum of j independent and identically distributed random variables), derive the mean and variance of N_j as

$$E(N_j) = \frac{j}{q} \; ; \; \mathrm{Var}(N_j) = \frac{j(1-q)}{q^2} \; .$$

1.5 Sums, transforms and limits

We have already encountered instances where the object of interest is a sum of independent random variables. The mean and the variance of such a sum can, of course, be obtained by adding together, respectively, the means and the variances of the constituent variables (the former can be done even without the assumption of independence). However, finding the distribution of the sum is, in general, a more difficult task which may involve a considerable computational effort.

Consider first the sum, S, of two independent discrete random variables, X and Y, whose values are non-negative integers. Denote by p_i, q_i and r_i ($i = 0, 1, \ldots$) the distributions of X, Y and S, respectively:

$$p_i = P(X = i) \; ; \; q_i = P(Y = i) \; ; \; r_i = P(S = i) \, . \qquad (1.68)$$

Since, in order for S to take the value i, X must take one of the values

1.5 Sums, transforms and limits

$0, 1, \ldots, i$ and Y must take the value $i - X$, we can write

$$\begin{aligned} r_i &= P(X+Y=i) = \sum_{j=0}^{i} P(X=j, Y=i-j) \\ &= \sum_{j=0}^{i} P(X=j)P(Y=i-j) = \sum_{j=0}^{i} p_j q_{i-j} \; ; \; i = 0, 1, \ldots \end{aligned} \quad (1.69)$$

The distribution r_i, obtained according to (1.69), is called the 'convolution' of the distributions p_i and q_i. Note that the reason for requiring X and Y to be independent is to ensure that the third equality in (1.69) holds. If X and Y are not independent, the distribution of their sum is not necessarily equal to the convolution of their distributions.

1.5.1 Generating functions

The convolution operation can be presented in another form, which is often much more convenient. With the distribution p_i is associated the function $p(z)$, defined as

$$p(z) = \sum_{i=0}^{\infty} p_i z^i . \quad (1.70)$$

This is called the 'generating function', or the 'z-transform', of the probabilities p_i. That function satisfies the relation $p(1) = 1$, since all the p_is must sum up to 1. Therefore, $p(z)$ is finite at least for real values z in the interval $-1 \leq z \leq 1$. The term 'generating function' is justified by the fact that if $p(z)$ is known, then the probabilities p_i are determined uniquely:

$$p_i = \frac{p^{(i)}(0)}{i!} \; ; \; i = 0, 1, \ldots , \quad (1.71)$$

where $p^{(i)}(0)$ is the ith derivative of $p(z)$ at $z = 0$.

The generating function can also be expressed in the form of an expectation:

$$p(z) = E(z^X) . \quad (1.72)$$

This follows directly from the definition of expectation, (1.40). It suffices to note that the random variable z^X takes the value z^i with probability p_i ($i = 0, 1, \ldots$).

Now introduce also the generating functions, $q(z)$ and $r(z)$, of the distributions q_i and r_i, respectively. These are defined by power series of the type (1.70), and can be expressed in the form (1.72). We can then write

$$\begin{aligned} r(z) &= E(z^S) = E(z^{X+Y}) = E(z^X z^Y) \\ &= E(z^X)E(z^Y) = p(z)q(z) \, . \end{aligned} \qquad (1.73)$$

Thus, the generating function of the convolution of two discrete distributions is equal to the product of their generating functions. Moreover, (1.73) generalizes in an obvious way to sums of any finite number of independent discrete random variables.

There are other important applications for generating functions, besides computing convolutions. For instance, we shall encounter problems where a discrete distribution has to be determined by solving an infinite system of linear equations. An elegant way of dealing with such a system is to reduce it to a single equation for an unknown generating function, and then solve the latter.

Given a generating function, it is quite easy to find the mean, and higher moments, of the corresponding random variable. Indeed, differentiating (1.72) with respect to z yields

$$p'(z) = E(Xz^{X-1}) \, . \qquad (1.74)$$

Hence, the expectation of X is obtained by setting $z = 1$ in (1.74):

$$p'(1) = E(X) \, . \qquad (1.75)$$

Similarly, the second derivative of $p(z)$ at $z = 1$ gives

$$p''(1) = E[X(X-1)] = E(X^2) - E(X) \, , \qquad (1.76)$$

which can be used to determine $E(X^2)$. In general,

$$p^{(n)}(1) = E[X(X-1)\ldots(X-n+1)] \, . \qquad (1.77)$$

The expectation in the right-hand side of (1.77) is sometimes referred to as the 'factorial moment' of X, of order n.

Examples

1. We saw that a single trial in a Bernoulli trials experiment is associated with a random variable, X, which takes value 0 with probability $1-q$ and 1 with probability q ($0 < q < 1$). The generating function of this distribution is $q(z) = 1-q+qz$. Therefore, the sum, S, of n independent such variables has a generating function given by

$$r(z) = (1-q+qz)^n \, . \qquad (1.78)$$

1.5 Sums, transforms and limits

This, according to (1.63), is the generating function of the binomial distribution. Indeed, if the right-hand side of (1.78) is expanded in powers of z, the coefficient of z^j turns out to be exactly the right-hand side of (1.59). Expressions (1.64) and (1.65) can be obtained by taking derivatives in (1.78) at $z = 1$.

2. The generating function of the geometric distribution is obtained as

$$p(z) = \sum_{i=1}^{\infty} q(1-q)^{i-1} z^i = \frac{qz}{1-(1-q)z} \,. \tag{1.79}$$

From here one could derive the mean, (1.56), and higher moments. Thus, $p'(1) = 1/q$. It also follows that the generating function of the negative binomial distribution (see exercise 4 of section 1.4) must be equal to $[p(z)]^j$, since that distribution is the j-fold convolution of the geometric one.

* * *

Let us now return to the distribution of the sum of independent random variables, this time in the case when the latter are continuous and have probability density functions. Let X and Y be two such random variables, with pdfs $f(x)$ and $g(x)$ respectively. Assume for the moment that X and Y take non-negative values. Then we can write an expression analogous to (1.69) for the pdf, $h(x)$, of the sum $S = X + Y$:

$$h(x) = \int_0^x f(t)g(x-t)dt \,. \tag{1.80}$$

This is justified by arguing that, in order for S to take the value x (which happens with probability $h(x)dx$), X must take some value $t < x$ (probability $f(t)dt$) and Y must be equal to $x-t$ (probability $g(x-t)dx$). Summing over all possible values for t gives the integral in the right-hand side of (1.80). That integral is also called the 'convolution' of the two functions f and g.

1.5.2 Laplace transforms and characteristic functions

Just as in the case of discrete distributions, manipulations with probability density functions are often simplified considerably by the introduction of transforms. When dealing with non-negatively valued random variables, it is convenient to associate with a probability density function, $f(x)$, its 'Laplace transform', $f^*(s)$, defined as

$$f^*(s) = \int_0^\infty e^{-sx} f(x) dx .\tag{1.81}$$

The right-hand side of (1.81) is finite for all $s \geq 0$ (and in fact for all complex s whose real part is non-negative), and satisfies the normalizing condition $f^*(0) = 1$. It can be demonstrated that the Laplace transform $f^*(s)$ determines the probability density function $f(x)$ uniquely. Indeed, there is an explicit formula expressing f in terms of f^*, which may be used for a numerical inversion:

$$f(x) = \frac{1}{2\pi} \int_{-\infty}^\infty e^{-isx} f^*(-is) ds ,\tag{1.82}$$

where $i = \sqrt{-1}$ is the imaginary unit.

Just like the generating function, the Laplace transform can be written in the form of an expectation:

$$f^*(s) = E(e^{-sX}) .\tag{1.83}$$

From this, it follows immediately that the Laplace transform of the sum of two independent random variables is equal to the product of their Laplace transforms:

$$h^*(s) = E(e^{-s(X+Y)}) = E(e^{-sX})E(e^{-sY}) = f^*(s)g^*(s) .\tag{1.84}$$

One can also obtain this result directly from (1.80). Note that, although the independence of X and Y is sufficient to ensure the validity of (1.84), it is not necessary. There are examples of random variables for which (1.84) holds, despite the fact that they are dependent on each other.

The derivatives of a Laplace transform at point $s = 0$ yield the moments of the corresponding random variable. For instance, differentiating (1.83) with respect to s we get

$$f^{*\prime}(s) = E(-Xe^{-sX}) .\tag{1.85}$$

Hence, setting $s = 0$,

$$E(X) = -f^{*\prime}(0) .\tag{1.86}$$

In general, if the nth moment of X exists, then it is given by

$$M_n = E(X^n) = (-1)^n f^{*(n)}(0) \; ; \; n = 1, 2, \ldots .\tag{1.87}$$

It may not be convenient, or even possible, to use Laplace transforms

1.5 Sums, transforms and limits

for random variables which take both positive and negative values. The integral in the right-hand side of (1.81), when taken over the interval $(-\infty, \infty)$, may diverge. To get around this difficulty, another transform of a similar type has been introduced. This is called the 'characteristic function', $\varphi_X(s)$, of the random variable X, and is defined as follows:

$$\varphi_X(s) = E(e^{isX}) \; ; \; -\infty < s < \infty \,. \tag{1.88}$$

Here, depending on whether X is discrete or continuous, the expectation in the right-hand side is given by a sum or an integral. The advantage of this definition is that, since e^{isX} is bounded, the characteristic function $\varphi_X(s)$ exists for all X (see exercise 2). The disadvantage, such as it is, lies in having to deal with complex numbers.

Like the other transforms that we have seen, the characteristic function of a random variable determines its distribution uniquely. Also, if X and Y are independent, then

$$\varphi_{X+Y}(s) = \varphi_X(s)\varphi_Y(s) \,. \tag{1.89}$$

If the nth moment of X exists, then it can be obtained from

$$M_n = E(X^n) = (-i)^n \varphi_X^{(n)}(0) \; ; \; n = 1, 2, \ldots \,. \tag{1.90}$$

Another direct consequence of the definition (1.88) is that, if Y is obtained from X by a linear transformation, i.e. $Y = aX + b$, then

$$\varphi_Y(s) = e^{isb}\varphi_X(as) \,. \tag{1.91}$$

In the modelling applications that will concern us in this book, the random variables of interest will almost always be non-negatively valued. Therefore, when we need to use transform methods, we shall be able to restrict ourselves to generating functions and Laplace transforms. However, before we leave this section, it will be instructive to examine a random variable for which the use of the characteristic function is indicated. At the same time, we shall introduce a distribution that plays a very important role in probability theory.

1.5.3 The normal distribution and central limit theorem

Let Z be a random variable whose probability density function is given by

$$f(x) = \frac{1}{\sqrt{2\pi}} e^{-x^2/2} \; ; \; -\infty < x < \infty \,. \tag{1.92}$$

This is called the 'standard normal' density. The corresponding distribution function is

$$F(x) = \frac{1}{\sqrt{2\pi}} \int_{-\infty}^{x} e^{-u^2/2} du \ . \tag{1.93}$$

The characteristic function of the standard normal random variable is equal to

$$\varphi(s) = \frac{1}{\sqrt{2\pi}} \int_{-\infty}^{\infty} e^{isx - x^2/2} dx = e^{-s^2/2} \ . \tag{1.94}$$

(The derivation of this result will be omitted: it requires contour integration and an application of Cauchy's theorem.)

The mean and variance of Z are immediately obtained by taking derivatives in (1.94) at $s = 0$:

$$E(Z) = (-i)\varphi'(0) = 0 \ ; \ \operatorname{Var}(X) = E(Z^2) = -\varphi''(0) = 1 \ . \tag{1.95}$$

Consider a linear transformation of the standard normal variable: $\mu + \sigma Z$, where $\sigma > 0$ and μ is an arbitrary real constant. The resulting random variable, which has mean μ and variance σ^2, or standard deviation σ, is said to have the general normal distribution. That variable will be denoted $Z(\mu, \sigma)$; thus, the standard normal variable is $Z(0, 1)$ (where there is no danger of ambiguity we shall continue to denote the latter by Z). The characteristic function of $Z(\mu, \sigma)$ is given by

$$\varphi(s) = e^{(i\mu s - \sigma^2 s^2/2)} \ . \tag{1.96}$$

It is easy to verify (see exercise 3) that the probability density function corresponding to (1.96) is

$$f(x) = \frac{1}{\sigma\sqrt{2\pi}} e^{-(x-\mu)^2/(2\sigma^2)} \ . \tag{1.97}$$

We thus have a family of normal distributions depending on two parameters, μ and σ. That family has the interesting property that it is closed with respect to convolution. In other words, if $Z(\mu_1, \sigma_1)$ and $Z(\mu_2, \sigma_2)$ are two independent normal random variables, then their sum is normally distributed. Since the variance of the sum is equal to the sum of the variances, we can write

$$Z(\mu_1, \sigma_1) + Z(\mu_2, \sigma_2) \sim Z(\mu_1 + \mu_2, \sqrt{\sigma_1^2 + \sigma_2^2}) \ , \tag{1.98}$$

where the symbol $\sim$ means 'equal in distribution'.

The above additive property is a direct consequence of (1.89) and

1.5 Sums, transforms and limits

(1.96): the product of two characteristic functions of type (1.96) is also of type (1.96).

It is equally easy to see that multiplying a normally distributed random variable with mean μ and variance σ^2 by an arbitrary real constant, c, produces a normally distributed random variable with mean $c\mu$ and variance $c^2\sigma^2$:

$$cZ(\mu,\sigma) \sim Z(c\mu, c\sigma) . \tag{1.99}$$

A simple corollary of the above is that if $Z_1, Z_2, \ldots, Z_n$ are independent standard normal random variables, then the random variable $Z_n = (Z_1 + Z_2 + \ldots + Z_n)/\sqrt{n}$ also has the standard normal distribution, no matter what the value of n:

$$\frac{1}{\sqrt{n}} \sum_{i=1}^{n} Z_i \sim Z \; ; \; n = 1, 2, \ldots . \tag{1.100}$$

Moreover, it turns out that the standard normal distribution is the only continuous distribution with this 'preservation' property.

Now consider a more general sum of independent and identically distributed random variables. Assume that $X_1, X_2, \ldots, X_n$ have mean 0 and variance 1, but otherwise allow their distribution to be arbitrary. Then it is possible to assert that the sum $Y_n = (X_1 + X_2 + \ldots + X_n)/\sqrt{n}$ has approximately the standard normal distribution. More precisely, the following result holds:

Theorem 1.1

$$\lim_{n \to \infty} \frac{1}{\sqrt{n}} \sum_{i=1}^{n} X_i \sim Z .$$

This is a special case of what is known as the 'central limit theorem'. The use of characteristic functions reduces its proof to a few uncomplicated manipulations.

Let $\varphi(s)$ be the characteristic function of X_i. Then the characteristic function of Y_n, $\psi_n(s)$, is given by

$$\psi_n(s) = [\varphi(s/\sqrt{n})]^n \; ; \; n = 1, 2, \ldots . \tag{1.101}$$

Now, since the first two moments of X_i exist and are equal to 0 and 1 respectively, one can write a partial Taylor expansion for $\varphi(s)$:

$$\begin{aligned} \varphi(s) &= \varphi(0) + \varphi'(0)s + \frac{1}{2}\varphi''(0)s^2 + o(s^2) \\ &= 1 - \frac{1}{2}s^2 + o(s^2) , \end{aligned} \tag{1.102}$$

where $o(x)$ is a function that tends to 0 faster than x, i.e. $o(x)/x \to 0$ when $x \to 0$.

Substituting (1.102) into (1.101) and letting $n \to \infty$, we get

$$\lim_{n \to \infty} \psi_n(s) = \lim_{n \to \infty} \left[1 - \frac{s^2}{2n} + o\left(\frac{s^2}{n}\right) \right]^n = e^{-s^2/2} . \qquad (1.103)$$

This establishes the theorem, since the right-hand side of (1.103) is precisely the characteristic function of the standard normal distribution.

Having got this result, it is easy to deal with the case when the X_is have arbitrary (but finite) mean μ and variance σ^2. Then it suffices to note that the random variables $(X_i - \mu)/\sigma$ have mean 0 and variance 1. Hence, the following normalized sum approaches the standard normal distribution:

$$\lim_{n \to \infty} \frac{1}{\sigma \sqrt{n}} \left(\sum_{i=1}^{n} X_i - n\mu \right) \sim Z . \qquad (1.104)$$

The central limit theorem is valid under considerably more general conditions than the ones stated here.

Example

3. Approximating the binomial distribution. Consider the number of successes, S_n, that occur in a sequence of n Bernoulli trials, where the probability of success is q. We have established already that this random variable has the binomial distribution, given by (1.59), with mean nq and variance $nq(1-q)$. Moreover, S_n can be represented as the sum of n independent indicator random variables, X_i, each of which has mean q and variance $q(1-q)$. This last fact, together with the central limit theorem, implies that a suitably normalized number of successes approaches the standard normal distribution:

$$\lim_{n \to \infty} \frac{S_n - nq}{\sqrt{nq(1-q)}} \sim Z .$$

This special case of (1.104) is known as the De Moivre–Laplace theorem.

One can therefore use the standard normal distribution, which is extensively tabulated, to approximate the binomial distribution when n is reasonably large. That is a useful facility, since the exact expressions appearing in (1.59) become progressively more difficult to compute, and more prone to numerical errors, as n increases.

Suppose, for instance, that we have a sample of 5000 items (e.g., machine parts), each of which is defective with probability 0.01, independently of the others. What is the probability, r, that the sample contains more than 60 defective items? The exact answer, according to (1.59), is given by:

$$r = P(S_{5000} > 60) = \sum_{j=61}^{5000} \binom{5000}{j} (0.01)^j (0.99)^{5000-j}.$$

The evaluation of this quantity is far from trivial. On the other hand, the random variable $(S_{5000} - 50)/\sqrt{49.5}$ is distributed approximately like the standard normal variable Z. Hence we easily obtain, using any table of the standard normal distribution,

$$P(S_{5000} > 60) \approx P\left(Z > \frac{60 - 50}{\sqrt{49.5}}\right) \approx P(Z > 1.42) \approx 0.08.$$

Exercises

1. In a Bernoulli trials experiment, let the number of trials, n, grow to infinity; at the same time, let the probability of success, q, decrease to 0 in such a way that the product $nq = \lambda$ remains constant. Show that the generating function of the number of successes, S_n, given by (1.78), approaches $e^{-\lambda(1-z)}$. Hence deduce that, in the limit, the probability that there will be k successes is equal to

$$p_k = \frac{\lambda^k}{k!} e^{-\lambda} \; ; \; k = 0, 1, \ldots .$$

This is the 'Poisson distribution', which we shall encounter again later.

2. Using the fact that the expectation operator satisfies the inequality $|E(Y)| \leq E(|Y|)$, show that the characteristic function of an arbitrary random variable satisfies

$$|\varphi(s)| \leq 1,$$

with equality at $s = 0$.

3. Show that the characteristic function of the general normal pdf in (1.97) is indeed given by (1.96): evaluate the integral

$$\int_{-\infty}^{\infty} e^{isx} f(x) dx \,,$$

by making a change of variable $u = (x - \mu)/\sigma$, and using (1.94).

4. Find the Laplace transforms of

(i) the uniform density $f(x) = 1/(b-a)$, $a \leq x \leq b$;
(ii) the 'exponential' density, $f(x) = \lambda e^{-\lambda x}$, $0 \leq x < \infty$ ($\lambda > 0$).

Hence obtain the first two moments of those distributions.

1.6 Literature

There is no lack of good books on probability theory. For the reader wishing a solid background in the subject, we would recommend the well-known book by Feller [2], a classic text. A shorter, more recent and very readable book by Whittle [6] is remarkable because it takes expectation as the fundamental concept, building the theory upon the axioms in section 1.3. An entertaining account of the history of the two approaches to probability theory can be found in an article by Stirzaker, [5].

For the more mathematically inclined readers, there are several books providing rigorous and comprehensive treatments of probability based on measure theory. Three widely used ones are Billingsley [1], Loeve [4] and Kingman and Taylor [3].

References

1. P. Billingsley, *Probability and Measure*, John Wiley & Sons, 1986.
2. W. Feller, *An Introduction to Probability Theory and Its Applications*, volume 1, John Wiley & Sons, 1968.
3. J.F.C. Kingman and S.J. Taylor, *Introduction to Measure and Probability*, Cambridge University Press, 1966.
4. M. Loeve, *Probability Theory*, Van Nostrand, 1955.
5. D. Stirzaker, "Probability Vicit Expectation", Chapter 1 in *Probability, Statistics and Optimization* (edited by F.P. Kelly), John Wiley & Sons, 1994.
6. P. Whittle, *Probability via Expectation*, Springer, 1992.

2

Arrivals and services

At a certain level of abstraction, computing and communication systems as well as banking, manufacturing and transport systems, can be described in terms of 'jobs' and 'servers', i.e. requests for service and devices that provide service. The jobs may be computing tasks, input/output commands, telephone calls, data packets. The servers may be processors, storage devices, communication channels, software modules. A model aimed at evaluating and predicting the performance of such a system has to capture the following essential aspects of its behaviour:

(a) The pattern of demand, i.e. the the manner in which jobs arrive into the system and the nature of services that they require.
(b) The competition for service, i.e. the effect of admission, queueing and routing policies on performance.

This chapter is devoted to (a). It introduces tools and results that are used when modelling the arrivals and services of jobs.

2.1 Renewal processes

Consider a phenomenon which takes place first at time 0 and thereafter keeps occurring, at random intervals, *ad infinitum*. Denote the consecutive instants of occurrence by T_n ($n = 0, 1, \ldots$; $T_0 = 0$), and let $S_n = T_n - T_{n-1}$ ($n = 1, 2, \ldots$) be the intervals between them. Assume that the random variables S_n are independent and identically distributed.

The sequence $\{T_n : n = 0, 1, \ldots\}$ is called a 'renewal process'. The instants T_n are referred to as the 'renewal points'; the intervals, S_n, between consecutive renewal points are the 'renewal intervals'. The name 'renewal' is justified by the fact that the process regenerates itself at every instant T_n. In other words, if the time origin is shifted to point T_n, for some $n \geq 1$, the subsequent process will be indistinguishable from the original.

Some of the applications of renewal processes are in the field of machine reliability. A new machine starts work at time 0. After a while it breaks down and is replaced (either immediately or after some delay) by an identical new machine; and so on. The renewal points are then the replacement instants, while a renewal interval consists of either the lifetime of a machine, or that lifetime plus the following replacement delay.

We shall use renewal processes mainly to model streams of jobs arriving into, or departing from, a system. These applications will also affect the terminology. For instance, we shall often refer to the renewal points and renewal intervals as 'arrival instants' and 'interarrival intervals', respectively.

A renewal process is completely characterized by specifying the common distribution function, $F(x)$, or the density function, $f(x)$ (if the latter exists), of the renewal intervals. The nth renewal point can be expressed as a sum of n renewal intervals:

$$T_n = \sum_{i=1}^{n} S_n \; ; \; n = 1, 2, \ldots . \tag{2.1}$$

Hence, the distribution function of T_n, $F_n(x)$, can be obtained by taking the n-fold convolution of $F(x)$ with itself (see section 1.5). That is, in general, a difficult operation. However, for large values of n, the central limit theorem tells us that T_n is approximately normally distributed. More precisely, if the mean and variance of the renewal intervals are m and v respectively, then T_n is distributed approximately like the normal random variable $Z(nm, \sqrt{nv})$. Alternatively, a normalized nth renewal instant has approximately the standard normal distribution:

$$\frac{T_n - nm}{\sqrt{nv}} \approx Z . \tag{2.2}$$

Now consider the number, K_t, of renewal points that fall in the interval $[0, t]$. Clearly, for that random variable to take value k, the renewal point T_{k-1} must occur before, or at time t, and T_k must occur after t. Therefore, the distribution of K_t is given by

$$P(K_t = k) = P(T_k \leq t < T_{k+1}) = P(T_{k-1} \leq t) - P(T_k \leq t)$$

$$= F_{k-1}(t) - F_k(t) \; ; \; k = 1, 2, \ldots \tag{2.3}$$

2.1 Renewal processes

where $F_n(t)$ is the distribution function of T_n and $F_0(t) = 1$ by definition. From here we can find the average number of renewals in the interval $[0, t]$. That average, which is of course a function of t, is called the 'renewal function' and is denoted by $H(t)$. There are two ways of determining the renewal function. From (2.3) and the definition of expectation it follows that:

$$H(t) = E(K_t) = \sum_{k=1}^{\infty} k[F_{k-1}(t) - F_k(t)] = \sum_{k=0}^{\infty} F_k(t) . \qquad (2.4)$$

Alternatively, one can condition the average number of renewals in $[0, t]$ upon the instant of the renewal point T_1: if the latter occurs after time t, then $H(t) = 1$; if T_1 occurs at time $t - x$, then $H(t)$ is equal to $1 + H(x)$. Hence

$$H(t) = 1 + \int_0^t f(t-x) H(x) dx , \qquad (2.5)$$

where $f(x)$ is the pdf of the renewal interval. This integral equation is sometimes referred to as the 'renewal equation'. If the renewal interval is a discrete random variable, then the integral in the right-hand side of (2.5) becomes a sum.

Thus, the renewal function can be found either by summing the infinite series (2.4), or by solving the renewal equation. Unfortunately, both those tasks are rather difficult to perform exactly. On the other hand, a very simple approximation is available when t is large. Intuitively, since the average length of the renewal interval is m, one can expect approximately t/m renewals during a long interval $[0, t]$. This is indeed true. In fact, a more accurate estimate can be made:

$$H(t) = \frac{t}{m} + \frac{v + m^2}{2m^2} + o(1) , \qquad (2.6)$$

where v is the variance of the renewal interval and $o(1) \to 0$ when $t \to \infty$. The proof of this result is non-trivial and will be omitted.

From (2.6) it follows that, *in the long run*, the average number of renewals during any interval of length y is approximately equal to

$$H(t+y) - H(t) \approx \frac{y}{m} . \qquad (2.7)$$

Suppose now that the distribution $F(x)$ of the renewal intervals is continuous, or at least that $F(0) = 0$. Then the probability of more than one renewal occurring in a small interval of length h is $o(h)$ (i.e. negligible compared to h). Hence, the average number of renewals in a small interval $(t, t+h)$, far away from the origin, is approximately equal to the probability that there is a renewal in that interval. Denoting that probability by $a(t, t+h)$ and applying (2.7), we conclude that when t is large,

$$a(t, t+h) = \frac{h}{m} + o(h) \ . \tag{2.8}$$

Note that the right-hand side of (2.8) does not depend on t.

In the context of job arrival streams, m is the average interval between consecutive arrival instants. The 'rate of arrivals', λ, can be defined either as the average number of arrivals per unit time, or as the limiting ratio $a(t, t+h)/h$ when $h \to 0$. Equations (2.7) and (2.8) can then be summarized in the following rather simple, but nevertheless important result:

Arrival rate lemma. In the long run, the rate of arrivals is constant and is equal to the reciprocal of the average interarrival interval:

$$\lambda = \frac{1}{m} \ .$$

2.1.1 Forward and backward renewal times

An quantity of considerable interest in connection with a renewal process is the so-called 'forward renewal time', also referred to as the 'residual life', or the 'random modification' of the renewal interval. This is defined as follows. Suppose that the renewal process is observed at random over a long run. In other words, an observation point is chosen so that it is equally likely to fall anywhere within a very long interval of time. Let that observation point be T, and the two renewal points between which it falls be T_{i-1} and T_i. The remainder of the observed renewal interval, $T_i - T$, is the forward renewal time, or the residual life of the renewal interval. The elapsed portion of the observed renewal interval, $T - T_{i-1}$, is called the 'backward renewal time' (figure 2.1).

In reliability applications, the residual life is the interval between the random observation point and the next machine replacement instant. In the case of an arrival stream, it is the interval between the observation point and the next arrival instant.

The following remark is prompted by the definition of 'random observation point': if the length of the observed renewal interval is y, then the observation point is uniformly distributed on the interval $(T_{i-1}, T_{i-1}+y)$. Hence, the probability density function of both the forward and backward renewal times, conditioned upon y, is $1/y$ (see section 1.2). Since

2.1 Renewal processes

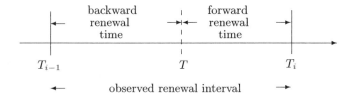

Fig. 2.1. Forward and backward renewal times

those two random variables are identically distributed we can concentrate on one of them, say the forward renewal time. However, it should be pointed out that the forward and backward renewal times are *not* independent of each other.

Denote the residual life by R and its (unconditional) probability density function by $r(x)$. Assume for the moment that the renewal intervals are discrete random variables, taking value x_i with probability $f(x_i)$ ($i = 1, 2, \ldots$).

Let $s(x_i)$ be the probability that the observed renewal interval is of length x_i. Note the distinction between $s(x_i)$ and $f(x_i)$. In general, the probability of observing a renewal interval of length x_i is not the same as the probability that an arbitrary renewal interval is of length x_i. The observation point is more likely to fall into a large renewal interval than into a small one, so the fact that a renewal interval is being observed distorts its distribution.

By conditioning upon the length of the observed renewal interval, the pdf of R can be expressed as

$$r(x) = \sum_{\{i \,:\, x_i > x\}} s(x_i) \frac{1}{x_i}. \qquad (2.9)$$

To determine $s(x_i)$, we argue as follows:

(a) According to the arrival rate lemma, the average number of renewal intervals per unit time is $1/m$;
(b) The average number of renewal intervals of length x_i per unit time is $f(x_i)/m$ (since each renewal interval is of length x_i with probability $f(x_i)$);
(c) The average fraction of time occupied by renewal intervals of length x_i is therefore equal to $x_i f(x_i)/m$;
(d) The probability of observing a renewal interval of length x_i is

equal to the average fraction of time occupied by intervals of length x_i, since the observation point is uniformly distributed over any time interval: $s(x_i) = x_i f(x_i)/m$.

Thus, the probability that the observed renewal interval is of length x_i is proportional to both x_i and the relative frequency, $f(x_i)$, with which intervals of length x_i occur.

Substituting the above expression into (2.9), we obtain

$$r(x) = \frac{1}{m} \sum_{\{i\,:\,x_i > x\}} f(x_i) = \frac{1 - F(x)}{m} \quad ; \quad x \geq 0 \,. \qquad (2.10)$$

The distribution function of the residual life, $F_R(x)$, is given by

$$F_R(x) = \frac{1}{m} \int_0^x [1 - F(x)]dx \,. \qquad (2.11)$$

If the renewal intervals are continuous random variables with a probability density function $f(x)$, then (2.9) becomes

$$r(x) = \int_x^\infty s(y) \frac{1}{y} dy \,. \qquad (2.12)$$

The pdf of the observed renewal interval, $s(x)$, is given by

$$s(x) = \frac{x f(x)}{m} \,. \qquad (2.13)$$

The final expression for $r(x)$ in (2.10), as well as (2.11), hold without change.

The average residual life is now easily calculated (see exercise 7 in section 1.3):

$$E(R) = \frac{1}{m} \int_0^\infty x[1 - F(x)]dx = \frac{M_2}{2m} \,, \qquad (2.14)$$

where M_2 is the second moment of the renewal interval. Again, this result holds for both discrete and continuous renewal intervals.

The approximation (2.6) can now be rewritten in terms of the residual life. Multiply (2.6) by m and substitute (2.14):

$$mH(t) = t + \frac{v + m^2}{2m} + o(1) = t + E(R) + o(1) \,.$$

In other words, the average number of renewals in $[0, t]$ (including the one at 0), multiplied by the average length of the renewal interval, gives approximately the time of the *next* renewal after point t. This is not unreasonable.

2.1 Renewal processes

2.1.2 The 'paradox' of residual life

The formula (2.14) for the average residual life leads to some unexpected conclusions. Intuition might suggest that, since the observation point is equally likely to fall anywhere, and since the average length of a renewal interval is m, the average residual life ought to be $m/2$. Instead, we learn from (2.14) that $E(R)$ is always greater than or equal to $m/2$, with equality only when $M_2 = m^2$, i.e. when the variance of the renewal intervals is 0. In fact, it may well happen that $E(R) > m$, i.e. the average residual life of the renewal interval can be greater than the average renewal interval! This apparent paradox is explained by the difference in distribution between the renewal interval and the observed renewal interval. The fact that a interval is observed makes it more likely to be large.

Examples

1. Irregular bus arrivals. Buses arrive at a certain stop according to a renewal process, with interarrival intervals taking two possible values: 95% of them are equal to 5 minutes, and 5% are equal to 60 minutes. Thus, the average interval between consecutive bus arrivals is

$$m = 5 \times 0.95 + 60 \times 0.05 = 4.75 + 3 = 7.75 \,.$$

How long, on the average, would a passenger coming at random to that stop have to wait for the next bus? To apply (2.14), we need the second moment of the interarrival interval:

$$M_2 = 5^2 \times 0.95 + 60^2 \times 0.05 = 23.75 + 180 = 203.75 \,.$$

Then we obtain

$$E(R) = \frac{203.75}{2 \times 7.75} \approx 13.15 \,.$$

So, buses arrive at average intervals of less than 8 minutes. Yet the customers do not wait for an average of less than 4 minutes; their average waiting time is more than 13 minutes.

2. Remaining service. Jobs are served, one after another, by a single server. Each service consists of two independent operations, performed sequentially. The durations of both operations are distributed uniformly, on the intervals $(0, 3)$ and $(0, 5)$, respectively. If this process is observed at random in the long run, what is the average interval until the completion of the current service?

The first and second moments of the renewal interval, which is now a sum of two independent random variables, $X_1 + X_2$, are given by

$$m = E(X_1) + E(X_2) = 1.5 + 2.5 = 4 .$$

$$M_2 = E(X_1^2) + E(X_2^2) + 2E(X_1)E(X_2) = 3 + \frac{25}{3} + 7.5 \approx 18.83 .$$

Hence, the average remaining service time is

$$E(R) \approx \frac{18.83}{8} \approx 2.35 .$$

Exercises

1. Jobs arrive into a system according to a renewal process, with interarrival intervals distributed uniformly on the interval $(0,2)$. Using the normal approximation, estimate the probability that there will be more than 50 arrivals within an interval of length 65.

2. Derive (2.14) by first showing that the average length of the observed renewal interval is M_2/m, and then arguing that the average residual life must be half of that quantity.

3. The lifetime of a machine is distributed according to the probability density function

$$f(x) = \frac{3}{(1+x)^4} \; ; \; x \geq 0 .$$

Assuming that the machine is replaced immediately after a breakdown, and that the resulting renewal process is observed at random, find (i) the average lifetime of a machine, (ii) the average lifetime of the observed machine and (iii) the average residual life of the observed machine.

4. Generalize expression (2.14) by showing that the nth moment of the residual life is given by

$$E(R^n) = \frac{M_{n+1}}{(n+1)m} ,$$

where M_n is the nth moment of the renewal interval.

2.2 The exponential distribution and its properties

A random variable, X, taking non-negative real values, is said to be 'exponentially distributed' if its distribution function has the form

$$F(x) = 1 - e^{-\lambda x} \; ; \; x \geq 0 . \qquad (2.15)$$

This function depends on a single parameter, $\lambda > 0$. The corresponding probability density function is

$$f(x) = \lambda e^{-\lambda x} \; ; \; x \geq 0 . \qquad (2.16)$$

Exponentially distributed random variables are commonly used to model random intervals of time whose lengths can be arbitrarily large: X may represent the service of a job, the duration of a communication session, the operative lifetime of a machine, the interval between consecutive arrivals, etc.

The first and second moments of X are

$$E(X) = \int_0^\infty x \lambda e^{-\lambda x} dx = \frac{1}{\lambda} \; ; \qquad (2.17)$$

$$M_2 = \int_0^\infty x^2 \lambda e^{-\lambda x} dx = \frac{2}{\lambda^2} . \qquad (2.18)$$

Consequently, the residual lifetime is equal to

$$E(R) = \frac{M_2}{2E(X)} = \frac{1}{\lambda} . \qquad (2.19)$$

So, the remainder of a randomly observed exponentially distributed interval has the same length, on the average, as a whole interval. Moreover, substituting (2.15) and (2.17) into (2.10), we find that the pdf of the remainder of the observed interval is the same as the pdf of a whole interval:

$$r(x) = \lambda e^{-\lambda x} = f(x) . \qquad (2.20)$$

The gist of the above observations is that the future progress of an exponentially distributed activity does not depend on its past; observing the activity to be in progress at a given point is the same as starting it at that point. This is known as the 'memoryless property' of the exponential distribution. Another way of stating that property is the following: the probability that the activity will continue for another interval of length at least x, given that it has already lasted time y, is independent of y:

$$P(X > y + x | X > y) = \frac{1 - F(y + x)}{1 - F(y)} = e^{-\lambda x} = P(X > x) . \qquad (2.21)$$

Thus, if interarrival intervals are distributed exponentially, the time until the next arrival is independent of the time since the last arrival. If the operative periods of a machine are distributed exponentially, the time to the next breakdown is independent of how long the machine has been working. If telephone calls have exponentially distributed lengths, the time until the line is vacated is independent of when the call started.

Once before we have encountered a memoryless property; that was in the case of the geometric distribution, considered in section 1.4.1. The analogy with that case extends further: just as the geometric is the only discrete distribution with the memoryless property, so the exponential is the only continuous distribution that has it. The proof of this uniqueness is outlined in exercise 2.

Completion rate. An immediate consequence of either (2.19) or (2.21) is that if the activity X is in progress at time t, then the probability that it will terminate in the interval $(t, t+h)$, where h is small, is independent of t and is approximately equal to λh:

$$P(R \leq h) = P(X \leq h) = 1 - e^{-\lambda h} = \lambda h + o(h) \ . \qquad (2.22)$$

In view of this result, the parameter λ is also called the 'completion rate' of the activity X.

2.2.1 First and last

Consider n independent activities whose durations, $X_1, X_2, \ldots, X_n$, are distributed exponentially with parameters $\lambda_1, \lambda_2, \ldots, \lambda_n$, respectively. Suppose that at time 0 all of them start in parallel. Denote by X_{min} the time when the first activity completes, and let $F_{min}(x)$ be its distribution function. The general expression obtained in section 1.2.4 implies that

$$F_{min}(x) = 1 - \prod_{i=1}^{n} e^{-\lambda_i x} = 1 - e^{-\lambda x} \ , \qquad (2.23)$$

where $\lambda = \lambda_1 + \lambda_2 + \ldots + \lambda_n$. The expectation of X_{min} is $1/\lambda$.

Thus, the shortest among several independent and exponentially distributed intervals is also distributed exponentially. Moreover, according to the memoryless property, if the n activities are observed to be in progress at any point in time, regardless of when each of them started, then the interval from the observation point until the first completion is distributed exponentially with parameter λ.

Let us find the probability, q_1, that activity X_1 is the first to complete.

2.2 The exponential distribution and its properties

Note that if X_1 (or its residual life) takes value x, then in order for it to complete first, the other activities must be longer than x. Integrating over all x we get

$$q_1 = \int_0^\infty \lambda_1 e^{-\lambda_1 x} \left(\prod_{j=2}^n e^{-\lambda_j x} \right) dx = \int_0^\infty \lambda_1 e^{-\lambda x} dx = \frac{\lambda_1}{\lambda}. \quad (2.24)$$

Similarly, the probability that X_i completes first is λ_i/λ ($i = 2, 3, \ldots, n$).

The exponential distribution arises naturally as a result of waiting for the first completion among a large number of independent activities. Indeed, there is a limit theorem stating that the interval until that first completion is approximately exponentially distributed, regardless of the distribution of each activity. We shall demonstrate this in the special case when $X_1, X_2, \ldots, X_n$ have the same distribution function, $F(x)$. Denote by Y_n the normalized first completion time when all activities start at time 0:

$$Y_n = n[\min(X_1, X_2, \ldots, X_n)] = \min(nX_1, nX_2, \ldots, nX_n). \quad (2.25)$$

Let $G_n(x)$ be the distribution function of Y_n. The normalization is introduced in order to ensure that Y_n does not shrink to 0 as n increases.

Theorem 2.1 *If $F(x)$ satisfies $F(0) = 0$ and $F'(0) = \beta > 0$, then in the limit $n \to \infty$, the distribution of the normalized first completion time becomes exponential with parameter β.*

The proof is quite straightforward. Since nX_i takes a value not exceeding x with probability $F(x/n)$, the expression for $G_n(x)$ is

$$G_n(x) = 1 - \left[1 - F(\frac{x}{n})\right]^n.$$

When n is large, x/n is close to 0 and so $F(x/n) = \beta x/n + o(1/n)$. Hence,

$$\lim_{n \to \infty} G_n(x) = \lim_{n \to \infty} \{1 - [1 - \frac{\beta x}{n} + o(\frac{1}{n})]^n\} = 1 - e^{-\beta x}. \qquad \text{qed.}$$

A similar result holds when a large number of independent and identically distributed activities are observed to be in progress at some random point in time. The role of Y_n is now played by the normalized interval, V_n, until the first completion after the observation point:

$$V_n = n[\min(R_1, R_2, \ldots, R_n)] = \min(nR_1, nR_2, \ldots, nR_n), \quad (2.26)$$

where R_i is the residual life of activity X_i.

Theorem 2.2 *In the limit $n \to \infty$, the distribution of V_n becomes exponential with parameter $1/m$, where $m = E(X_i)$.*

This theorem is established by applying the previous one to the residual lives, whose common distribution function is given by (2.11). The latter satisfies $F_R(0) = 0$ and $F'_R(0) = 1/m$.

The above results explain why certain random variables tend to be distributed exponentially. For instance, let Y be the lifetime of a complex device consisting of many vital components. Since the device stops working when the first of those components fails, the distribution of Y can be expected to be approximately exponential.

Now consider the interval until the completion of the *last* of n independent activities, all starting at time 0. The distribution function of that random variable, $F_{\max}(x)$, was found in section 1.2.4. In general it is difficult to work with that distribution; even computing the mean can be a hard task. However, if all variables X_i are exponentially distributed with the same parameter, λ, then the average interval until the nth completion, m_n, can be obtained rather simply as follows: the average time until the first completion is equal to $1/(n\lambda)$; after the first completion, there are $n-1$ activities in progress and, according to the memoryless property, the average remaining time until the last completion is m_{n-1}. In other words, $m_n = 1/(n\lambda) + m_{n-1}$. The solution of these recurrences is

$$m_n = \frac{1}{\lambda} \sum_{k=1}^{n} \frac{1}{k}. \tag{2.27}$$

Of course this expression continues to hold if the n activities do not all start at time 0, but are observed to be in progress at some random point. When n is large, m_n is approximately equal to $(1/\lambda) \ln n$.

Exercises

1. Find the variance and the coefficient of variation (defined in section 1.3.2) of the exponential distribution.

2. Show that if a non-negative random variable X has a pdf and satisfies (2.21), then it is exponentially distributed. Hint: rewrite (2.21) as $1 - F(x+y) = [1 - F(x)][1 - F(y)]$ and hence as $F(x+y) - F(x) = F(y)[1-F(x)]$; divide both sides by y and let $y \to 0$; show that the unique solution of the resulting differential equation, subject to $F(0) = 0$ and $F'(0) > 0$, is the exponential distribution function.

3. The random variables $X_1, X_2, \ldots, X_n$, are independent and exponentially distributed with parameters $\lambda_1, \lambda_2, \ldots, \lambda_n$, respectively. Find the conditional pdf of X_1, given that it is the smallest of the n variables. Hint: the probability that X_1 takes value x, given that it is the smallest, is equal to the probability that X_1 takes value x *and* the other variables exceed x, divided by the probability that X_1 is the smallest.

4. Show that the Laplace transform, $f^*(s)$, of the exponential density with parameter λ (see section 1.5.2) is equal to

$$f^*(s) = \frac{\lambda}{\lambda + s}.$$

5. Let N be an integer-valued random variable distributed geometrically with parameter q: $P(N = n) = (1-q)^{n-1} q$. Let also $X_1, X_2, \ldots$ be independent random variables, all distributed exponentially with parameter λ. Show that the random sum

$$X = \sum_{i=1}^{N} X_i,$$

is distributed exponentially with parameter $q\lambda$. (Hint: condition upon the value of N and use the Laplace transform derived in exercise 4).

6. The random variables $X_1, X_2, \ldots, X_n$, are independent and uniformly distributed on the interval $(0,1)$. Find the distribution and the mean of the smallest (not normalized) and the largest among them.

2.3 The Poisson process

A renewal process with exponentially distributed renewal intervals is called a 'Poisson process'. The parameter, λ, of the exponential distribution is called the 'rate' of the Poisson process for reasons which will become obvious shortly. As we shall see, these processes are quite remarkable in many ways, and are very widely used to model streams of arrivals. The limiting property of the exponential distribution makes the Poisson process a reasonable approximation of the request arrival instants in many computing and communication systems. On the other hand, the memoryless property facilitates considerably the analysis of the resulting models.

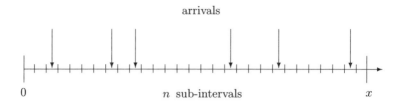

Fig. 2.2. Poisson arrivals in an interval of length x

According to (2.22), if a Poisson arrival process is observed at a random point in time, t, an arrival will occur within the interval $(t, t + h)$ (i.e. the current interarrival interval will terminate) with probability $\lambda h + o(h)$. This agrees with the arrival rate lemma of section 2.1 and with (2.8), and justifies referring to λ as the arrival rate. However, whereas for a general renewal process (2.8) is valid only in the long run (for large values of t), in the case of a Poisson process it holds at all times. In fact, it is possible to take this as a fundamental characteristic and define the Poisson process as an arrival process which satisfies the following conditions:

1. The probability of an arrival in $(t, t + h)$ is equal to $\lambda h + o(h)$, regardless of t and the process history before t;
2. The probability of more than one arrival in $(t, t + h)$ is $o(h)$, regardless of t and the process history before t.

The fact that the interarrival intervals are exponentially distributed with parameter λ would then follow from this definition.

From the memoryless property it follows that the number of arrivals during any finite interval, $(t, t+x)$, depends only on its length, x, and not on its position. Denote the random variable representing that number by K_x (note that the interval is open—it does not begin with an arrival as in section 2.1). The distribution of K_x can be derived as follows.

Divide the interval of length x into a large number, n, of sub-intervals; the length of each sub-interval, $h = x/n$, approaches 0 when $n \to \infty$ (figure 2.2).

When h is very small, the probability that there is more than one arrival in one sub-interval is negligible. Then the events $\{K_x = k\}$ and $\{k$ sub-intervals contain an arrival$\}$, can be treated as equivalent. Moreover, the occurrence of an arrival in any sub-interval is independent of what happens in the other sub-intervals: these occurrences can be considered

2.3 The Poisson process

as successes in a sequence of n Bernoulli trials. Therefore, K_x has approximately the binomial distribution (section 1.4.2), with probability of success λh:

$$p_k(x) = P(K_x = k) \approx \binom{n}{k} (\lambda h)^k (1 - \lambda h)^{n-k}.$$

Replacing h with x/n and rearranging terms, this can be re-written as

$$p_k(x) \approx \frac{(\lambda x)^k}{k!} \left(1 - \frac{\lambda x}{n}\right)^n \left(1 - \frac{\lambda x}{n}\right)^{-k} \frac{n(n-1)\ldots(n-k+1)}{n^k}.$$

Now let $n \to \infty$, keeping k fixed. In the limit, the approximate equality will become exact. The first term in the right-hand side does not depend on n, the second approaches $e^{-\lambda x}$, while the third and fourth approach 1. Thus we find that the distribution of K_x is given by

$$p_k(x) = \frac{(\lambda x)^k}{k!} e^{-\lambda x} \quad ; \quad k = 0, 1, \ldots \quad . \tag{2.28}$$

This is known as the the 'Poisson distribution'. Note that although the above derivation involved two constants—the arrival rate λ and the interval length x—the Poisson distribution depends on a single parameter: the product λx.

The average number of arrivals in an interval of length x is obtained from

$$E(K_x) = \sum_{k=1}^{\infty} k p_k(x) = \lambda x \, e^{-\lambda x} \sum_{k=1}^{\infty} \frac{(\lambda x)^{k-1}}{(k-1)!} = \lambda x \, . \tag{2.29}$$

Hence the arrival rate λ can also be interpreted as the average number of arrivals per unit time. Again, it is notable that (2.29) is an exact result, whereas the analogous expression for general renewal processes, (2.7), is valid only in the long run.

2.3.1 Properties of the Poisson process

Let us forget for the moment the physical interpretation of the Poisson process and just consider the class of Poisson distributions depending on one parameter, β. That class has the useful property of being closed with respect to convolution. In other words, the following is true:

Theorem 2.3 *If K_1 and K_2 are two independent Poisson random variables with parameters β_1 and β_2 respectively, then the sum $K = K_1 + K_2$ has the Poisson distribution with parameter $\beta_1 + \beta_2$.*

Perhaps the simplest way to prove this result is by introducing the generating function of the Poisson distribution with parameter β (see section 1.5.1):

$$g(z) = \sum_{k=0}^{\infty} \frac{\beta^k}{k!} e^{-\beta} z^k = e^{-\beta(1-z)}. \tag{2.30}$$

Then, denoting the generating functions corresponding to K_1, K_2 and K by $g_1(z)$, $g_2(z)$ and $h(z)$, respectively, we get

$$h(z) = g_1(z)g_2(z) = e^{-\beta_1(1-z)} e^{-\beta_2(1-z)} = e^{-(\beta_1+\beta_2)(1-z)},$$

which is the generating function of a Poisson random variable with parameter $\beta_1 + \beta_2$, qed.

The following two properties of the Poisson process are direct consequences of the the above theorem.

The union property. If I_1 and I_2 are two disjoint intervals of lengths x_1 and x_2 respectively, and I is their union, then the number of arrivals in I from a Poisson process with rate λ has the Poisson distribution with parameter $\lambda(x_1 + x_2)$.

Apply theorem 2.3 to the numbers of arrivals in I_1 and I_2, noting that, because of the memoryless property, they are independent of each other. Obviously, the argument extends to any finite number of disjoint intervals.

The union property can also be taken as the definition of a Poisson process. It implies both the exponential distribution of the interarrival intervals and their independence.

The superposition property. If B_1 and B_2 are two independent Poisson processes with rates λ_1 and λ_2 respectively (e.g. two groups of users submitting jobs to a central computer), then the the total number of arrivals from B_1 and B_2 during an interval of length x has the Poisson distribution with parameter $(\lambda_1 + \lambda_2)x$.

Apply theorem 2.3 to the numbers of arrivals from B_1 and B_2. Again, the argument extends to the superposition of any finite number of independent Poisson processes.

The normal approximation. The union and superposition properties suggest that the Poisson distribution approaches the normal one when the parameter λx increases. Indeed, it is intuitively clear that the number of arrivals during a long interval of time, or from a process with a large arrival rate (or both), can be regarded as a sum of the numbers of arrivals during many sub-intervals of finite length, or from many sub-processes

2.3 The Poisson process

with finite rates. Such a sum would obey the central limit theorem and would be approximately normally distributed. This proposition can be proved rigorously.

More precisely, if the random variable K has the Poisson distribution with mean λx (and variance λx; see exercise 1), then the normalized random variable $(K - \lambda x)/\sqrt{\lambda x}$ approaches in distribution the standard normal variable, Z, when $\lambda x \to \infty$.

Example

1. The normal approximation can simplify considerably the computational task involved in evaluating Poisson probabilities. Consider a computing system where jobs arrive according to a Poisson process, at an average rate of 2 jobs per second. Suppose that we are interested in the number of jobs, K, that arrive during a two-minute interval, and wish to find the probability that that number does not exceed 260. Here we have $\lambda = 2$, $x = 120$ and $\lambda x = 240$. The exact answer to this question is given by the expression

$$P(K \leq 260) = \sum_{k=0}^{260} \frac{240^k}{k!} e^{-240},$$

which does not lend itself easily to numerical evaluation.

Using the normal approximation, we get

$$P(K \leq 260) = P\left(\frac{K - 240}{\sqrt{240}} \leq \frac{20}{\sqrt{240}}\right) \approx P(Z \leq 1.29) \approx 0.90.$$

For practical purposes, the normal approximation can be applied successfully if λx is greater than about 20.

* * *

Consider now the operation of splitting, or decomposing, a Poisson arrival process A, with rate λ, into two arrival processes, B_1 and B_2. For instance, a computer centre may have two processors capable of executing the jobs submitted by its users; some of the incoming jobs could be directed to the first processor and others to the second.

The decomposition is performed by a sequence of Bernoulli trials: every arrival of the process A is assigned to the process B_1 with probability q, and to B_2 with probability $1 - q$, regardless of all previous assignments. Let K, K_1 and K_2 be the numbers of arrivals associated

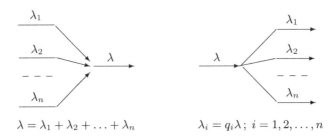

Fig. 2.3. Merging and splitting of Poisson arrivals

with A, B_1 and B_2 respectively, during an interval (or a set of disjoint intervals) of length x. We know that K, which is equal to $K_1 + K_2$, has the Poisson distribution with parameter λx; the problem is to determine the joint distribution of K_1 and K_2. Bearing in mind that, if the value of K is given, then K_1 and K_2 are the numbers of successes and failures in the corresponding set of Bernoulli trials, we can write

$$\begin{aligned} P(K_1 = i, K_2 = j) &= P(K_1 = i, K_2 = j \mid K = i+j) P(K = i+j) \\ &= \binom{i+j}{i} q^i (1-q)^j \frac{(\lambda x)^{i+j}}{(i+j)!} e^{-\lambda x} \\ &= \frac{(q\lambda x)^i}{i!} e^{-q\lambda x} \frac{[(1-q)\lambda x]^j}{j!} e^{-(1-q)\lambda x} \ . \end{aligned} \quad (2.31)$$

This equation implies that the distributions of K_1 and K_2 are Poisson, with parameters $q\lambda x$ and $(1-q)\lambda x$ respectively. Moreover, those random variables are independent of each other. An obvious generalization of the above argument leads to:

The decomposition property. If a Poisson process, A, with rate λ, is decomposed into processes $B_1, B_2, \ldots, B_n$, by assigning each arrival in A to B_i with probability q_i ($q_1 + q_2 + \ldots + q_n = 1$), independently of all previous assignments, then $B_1, B_2, \ldots, B_n$ are Poisson processes with rates $q_1\lambda$, $q_2\lambda$, ..., $q_n\lambda$ respectively, and are independent of each other.

The superposition and decomposition of Poisson processes are illustrated in figure 2.3.

When studying the behaviour of a system where jobs arrive at random, we shall sometimes wish to look at its state 'through the eyes' of the new arrivals. To be more precise, suppose that the system state at time t is described by the random variable (or vector of random variables) S_t. If t is a moment when an arrival occurs, then the state just before that

2.3 The Poisson process

moment, S_{t^-}, is the state 'seen' by the arriving job. Now, obviously the fact that there is an arrival at t influences S_t and the subsequent states, S_y, $y > t$. However, does that fact influence the state seen by the new arrival, S_{t^-}?

In general, it does. However, if jobs arrive into the system according to a Poisson process, then for large values of t, the distribution of S_{t^-} is independent of whether there is an arrival at t. This is usually stated as follows:

The PASTA property (Poisson Arrivals See Time Averages). In the long-term, the system state seen by an arrival from a Poisson process has the same distribution as the state seen by a 'random observer', i.e. at a point in time chosen at random.

We offer the following intuitive justification of PASTA. The system state seen at time t, S_{t^-}, is influenced only by the arrivals that occur before t. When the arrival process is Poisson, the interarrival intervals are exponentially distributed and have the memoryless property, looking both forward and backward in time. Hence, the interval between t and the nearest preceding arrival (the backward renewal time from section 2.1.1) has the same distribution regardless of whether t is an arrival instant. Consequently, the fact that a job arrives at time t has no implications about previous arrivals. Therefore, the incoming job and a random observer see the same distribution of the system state.

A non-PASTA example

2. The PASTA property may be appreciated better by considering a system where jobs do not arrive according to a Poisson process and see a markedly different distribution of the system state from that seen by a random observer. In a manufacturing plant, machine parts arrive on a conveyor belt to a robot operator, at intervals which are uniformly distributed between 40 and 60 seconds. The operator works on each part for exactly 30 seconds. This system, which is illustrated in figure 2.4, is in one of two possible states: 0 if the robot is idle, 1 if busy. Denote by p_1 and $\tilde{p}_1$ the long-term probabilities that a random observer and an arriving part see state 1, respectively.

Clearly, every arriving part sees an idle operator and so $\tilde{p}_1 = 0$. On the other hand, a random observer may well find the operator busy. Out of an average interarrival interval of 50 seconds, the operator is busy for 30 seconds; hence $p_1 = 3/5$ (this can also be derived formally by conditioning upon the length of the observed interarrival interval, x,

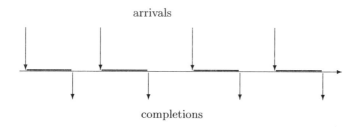

Fig. 2.4. Conveyor belt operations

whose pdf is provided by (2.13); given x, the probability of seeing state 1 is $30/x$).

If the arrival process is Poisson with the same rate, i.e. if the interarrival intervals are exponentially distributed with mean 50 seconds, then in the long run, both incoming parts and random observer will see state 1 with the same probability. The results presented in the next chapter indicate that that probability remains $3/5$.

* * *

Given all the nice properties of the Poisson process, it is hardly surprising that the analysis of a system model is greatly facilitated when one can assume that the stream of demands arriving into the system is Poisson. Fortunately, such an assumption can often be justified on grounds other than those of analytical convenience.

The Poisson process as a limit. The Poisson distribution, like the normal one, turns up as the limiting distribution of large sums. The following result holds under some mild assumptions:

Theorem 2.4 *The superposition of n arbitrary, independent and identically distributed renewal processes, each with an average renewal interval n/λ, approaches a Poisson process with rate λ when $n \to \infty$.*

This theorem follows from theorem 2.2 in section 2.2.1. The interval between a random observation point and the next arrival instant in the merged process is distributed like the variable V_n—the first completion among n normalized residual renewal intervals. That distribution approaches the exponential when $n \to \infty$.

2.3 The Poisson process

The superposition limit theorem applies even more generally. The processes being merged do not have to be identically distributed, as long as a few of them do not dominate all the others.

In view of this result, it can be argued that if the demands arriving into a system are submitted by many independent users, and if each user's requests are reasonably widely spaced, then the overall arrival process is approximately Poisson.

2.3.2 Arrivals during a random interval

It is often desirable to know the distribution of the number of arrivals, K_X, from a Poisson process during an interval whose length, X, is a random variable. Suppose that the probability density function of X is $f(x)$. Then the probability that there are k arrivals during the interval, p_k, can be obtained by conditioning upon its length and integrating:

$$p_k = P(K_X = k) = \int_0^\infty \frac{(\lambda x)^k}{k!} e^{-\lambda x} f(x) dx \; ; \; k = 0, 1, \ldots . \quad (2.32)$$

These expressions cannot, in general, be reduced to a simpler form. However, the generating function, $g(z)$, of the probabilities p_k, is very elegantly related to the Laplace transform, $f^*(s)$, of the density $f(x)$. Remember the definitions (sections 1.5.1 and 1.5.2):

$$g(z) = \sum_{k=0}^\infty p_k z^k \; ; \; f^*(s) = \int_0^\infty e^{-sx} f(x) dx . \quad (2.33)$$

Substituting (2.32) into (2.33) and changing the order of summation and integration, we obtain

$$\begin{aligned} g(z) &= \int_0^\infty \left[\sum_{k=0}^\infty \frac{(\lambda x)^k}{k!} z^k \right] e^{-\lambda x} f(x) dx = \int_0^\infty e^{-(\lambda - \lambda z)x} f(x) dx \\ &= f^*(\lambda - \lambda z) . \end{aligned} \quad (2.34)$$

This is a very useful formula. All moments of K_X can be determined by taking derivatives in (2.34) at $z = 1$. For instance, the average number of arrivals during X is

$$E(K_X) = g'(1) = -\lambda f^{*'}(0) = \lambda b , \quad (2.35)$$

where b is the average length of the interval.

If the random variable X is discrete, the integral in (2.34) will be a sum. The final expression for $g(z)$ can be written in a form which holds regardless of the distribution of X:

$$g(z) = E[e^{-(\lambda - \lambda z)X}]. \qquad (2.36)$$

Exercises

1. Show, by taking derivatives of the generating function (2.30), that the second moment of a Poisson random variable with parameter λx is $(\lambda x)^2 + \lambda x$. Hence, the variance of the number of arrivals during an interval of length x is equal to λx.

2. A certain computer system is subject to crashes which occur according to a Poisson process, at the rate of 10 per year (assume, for simplicity, that the repairs are instantaneous). Given that there was exactly one crash during the month of January, what is the probability that it occurred on the last day of that month?

3. (Generalization of the previous exercise.) Let K_x be the number of arrivals from a Poisson process with rate λ during an interval of length x. Consider a sub-interval of length y ($y < x$), and let K_y be the number of arrivals that fall within it. Show that the conditional distribution of K_y, given that $K_x = n$, is binomial:

$$P(K_y = k \mid K_x = n) = \binom{n}{k} q^k (1-q)^{n-k},$$

where $q = y/x$. Note that this does not depend on the position of the smaller interval within the larger one; only the lengths matter. Hence conclude that, given the number of arrivals in an interval, the arrival points are uniformly distributed on that interval, and are independent of each other.

4. Compilation jobs arrive at a computer centre according to a Poisson process, at the rate of 3 jobs per minute. The compiler requested in each case is C++ with probability 0.6, C with probability 0.3 and Pascal with probability 0.1. What is the probability that the numbers of C++, C and Pascal compilations requested during an interval of one hour will exceed 100, 50 and 20, respectively? Hint: use the decomposition property and the normal approximation.

2.4 Application: ALOHA and CSMA

5. Let T_n be the instant of the nth arrival in a Poisson process with rate λ. This is the sum of n independent random variables, each distributed exponentially with parameter λ. Show that the distribution function of T_n, $F_n(x)$, is equal to

$$F_n(x) = 1 - \sum_{k=0}^{n-1} \frac{(\lambda x)^k}{k!} e^{-\lambda x}.$$

This is called the 'Erlang distribution function'. The probability density function corresponding to it is

$$f_n(x) = F'_n(x) = \frac{\lambda (\lambda x)^{n-1} e^{-\lambda x}}{(n-1)!}.$$

(Hint: use the fact that $P(T_n \leq x) = P(K_x \geq n)$.)

6. Cars arrive at a police check point in a Poisson stream with rate λ. Each car has inadequate brakes with probability q, independently of all the others. What is the distribution of the interarrival interval for cars with inadequate brakes?

7. Find the second moment, $E(K_X^2)$, of the number of arrivals from a Poisson process with rate λ, during a random interval whose mean and second moment are b and M_2, respectively.

8. Derive an explicit formula for the generating function of the number of arrivals from a Poisson process with rate λ, during a random interval whose length is distributed exponentially with parameter μ. (Hint: see exercise 4 in section 2.2).

2.4 Application: ALOHA and CSMA

A famous packet switching communication network, called ALOHA, was developed at the University of Hawaii in the early 1970s. That system is completely decentralized: the communication channel is a radio frequency and to transmit a packet is to broadcast it. The moment a packet is ready for transmission at any node, it is transmitted. All nodes listen to the channel all the time; a destination recognises the packets intended for it by an appropriate address field.

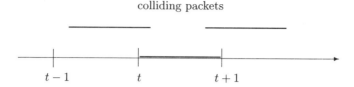

Fig. 2.5. Collisions of ALOHA packets

The price paid for the simplicity of this channel-sharing protocol is that not all transmissions are successful. If the broadcasts of two or more packets overlap, the information contained in those packets is corrupted. Such an event is called a 'collision'. When a collision occurs, the senders that are involved in it soon become aware of the fact (because they, too, are listening to the channel). All participants then back off and attempt to retransmit their packets after random periods of time. That randomness is essential: without it, a collision would necessarily be followed by another collision.

The first performance question that arises in connection with the ALOHA protocol concerns the expected maximum achievable throughput of the channel. Assume, for simplicity, that all packets are of fixed length and choose the time unit so that the packet transmission time is 1. Then, if there were no collisions and packets were transmitted one after the other without any gaps, the throughput would be 1 packet per unit time. How close to that ideal can one expect to get in reality?

The stream of arrivals at the channel consists of newly generated packets and old ones, offered for retransmission because of past collisions. Denote the total arrival rate by G (packets per unit time). Let S be the average number of successfully transmitted packets per unit time, i.e. the throughput. Then, if q is the probability that an incoming packet is transmitted successfully, we can write

$$S = Gq \,. \qquad (2.37)$$

In order to estimate the probability q, assume that the total arrival stream is Poisson, with rate G. This is not necessarily correct, but will do as a first approximation. Now, a packet arriving at time t will be successful if its transmission does not overlap with that of any other packet, i.e. if no other packets arrive in the interval $(t-1, t+1)$. This situation is illustrated in figure 2.5.

2.4 Application: ALOHA and CSMA

From the properties of the Poisson process it follows that the probability that there are no other arrivals during the interval $(t-1, t+1)$ is independent of t and is equal to $q = e^{-2G}$. Hence, the relation (2.37) between the offered traffic rate and the throughput becomes

$$S = Ge^{-2G}. \tag{2.38}$$

The right-hand side of (2.38) is initially an increasing function of G; it reaches a maximum and then decreases, approaching 0 when $G \to \infty$. To find the largest value that S can possibly attain, take the derivative of (2.38) with respect to G and equate it to 0. This immediately reveals that the optimal traffic rate is $G = 1/2$, and that the corresponding maximal throughput is

$$S_{\max} = \frac{1}{2e} \approx 0.184. \tag{2.39}$$

We have arrived at the somewhat disappointing conclusion that, under the free-for-all ALOHA protocol (also referred to as 'pure ALOHA'), one cannot expect to achieve more than about 18% of the ideal one-packet-per-unit-time throughput. Moreover, any attempt to get more packets through by increasing the offered traffic rate is bound to misfire eventually and result in a dramatic drop in throughput.

The poor utilization of the available channel capacity is due to packet collisions (and the consequent necessity for retransmission). It is natural, therefore, to attempt to improve matters by reducing the period of time during which collisions are possible. One way of achieving this is to allow transmissions to start only at selected points in time, say the integer points $\{0, 1, 2, \ldots\}$. Thus, if a packet arrives (becomes ready for transmission) sometime during the interval $(n-1, n)$, it will be transmitted during the interval $n, n+1$. That transmission will be successful if no other packets are transmitted during the same interval.

This modified protocol is referred to as 'slotted ALOHA'; the intervals $(n, n+1)$ $(n = 0, 1, \ldots)$ are called 'slots'. It is readily seen that the effect of slotting is to double the maximum achievable throughput. Indeed, a packet arriving in a given slot will avoid collision if, and only if, there are no other arrivals in the same slot. The probability of that event is $q = e^{-G}$. Thus, in the slotted ALOHA system, the throughput is related to the offered traffic rate as follows:

$$S = Ge^{-G}. \tag{2.40}$$

The function (2.40) is similar in shape to the one in (2.38). However, it

reaches a maximum of
$$S_{\max} = \frac{1}{e} \approx 0.37 , \qquad (2.41)$$
at $G = 1$. This is a two-fold increase, compared with pure ALOHA.

A further improvement in performance can be achieved by making a better use of the fact that all nodes are able to monitor the channel continuously. Suppose that a transmission can be started at any time but only if the sender, having listened to the channel, has found it idle. If another transmission is sensed to be in progress, the sender backs off and tries again after a random period of time. Such a protocol is commonly known as 'Carrier Sense Multiple Access', or CSMA. Perhaps the best-known example of a CSMA network is Ethernet, where the channel is a cable connecting all nodes.

Collisions may still occur under a CSMA protocol. Because of the finite speed at which signals are propagated, a sender may think that the channel is idle, whereas in fact another transmission has just started. Let d be the propagation delay of the channel, i.e. the time for the signal to travel from one station to another. If a transmission is attempted at time t, and another one starts at some point in the interval $(t-d, t+d)$, then the later of the two senders will not be aware of the earlier one, and the result will be a collision. Consequently, a transmission started in the belief that the channel is idle will be successful if, and only if, there are no other arrivals in an interval of length $2d$. That is, $q = e^{-2dG}$.

An approximate relation between the throughput, S, and the offered traffic rate, G, can be obtained as follows. The throughput is equal to the arrival rate of *successful* packets. Since the transmission time is 1, S is also equal to the probability, or the fraction of time, that the channel is busy with a successful transmission. According to the CSMA protocol, no one attempts to transmit during a successful transmission. During the rest of the time, such attempts are made at rate G. Therefore, the overall rate at which transmissions are started is $(1-S)G$. Remembering that each attempt is successful with probability e^{-2dG}, we can write

$$S = (1-S)Ge^{-2dG} . \qquad (2.42)$$

Solving this equation for S yields

$$S = \frac{Ge^{-2dG}}{1 + Ge^{-2dG}} . \qquad (2.43)$$

The function in the right-hand side of (2.43) reaches a maximum of $1/(1 + 2de)$, for $G = 1/(2d)$. This is a much better state of affairs than in the other two systems, since the value of d is usually quite small. Of course, if the offered traffic rate G is allowed to exceed the optimal value by much, the throughput will again drop to something near zero.

This raises the question of the stability of these protocols. Given the rate at which new packets are generated by the user population, what happens to the total offered traffic rate in the long run? Does it remain bounded and reasonably low, so that an acceptable throughput can be achieved, or does it keep growing, eventually bringing the throughput down to zero? To answer that question it is necessary to study the dynamic behaviour of the number of packets awaiting retransmission. We shall return to this topic in chapter 5, after the theory and tools of Markov chain modelling become available.

2.5 Literature

A broad coverage and extra results on renewal processes can be found in Cox [2], and also in Cox and Isham [3]. These sources discuss in some detail the limiting behaviour of the superposition of a large number of independent renewal processes. The exponential distribution and the Poisson process are of course mentioned in most books on probability theory and stochastic processes. A good example is Çinlar [1]. For an extensive treatment of computer networks, the reader is directed to Tanenbaum [5]. A more detailed analysis of CSMA performance than the one presented here can be found in Kleinrock and Tobagi [4].

References

1. E. Çinlar, *Introduction to Stochastic Processes*, Prentice-Hall, 1975.
2. D.R. Cox, *Renewal Theory*, Methuen, 1962.
3. D.R. Cox and V. Isham, *Point Processes*, Chapman and Hall, 1980.
4. L. Kleinrock and F.A. Tobagi, "Packet Switching in Radio Channels: Part 1—Carrier Sense Multiple Access Modes and Their Throughput-Delay Characteristics", *IEEE Trans. Comm.*, **23**, 12, 1975.
5. A.S. Tanenbaum, *Computer Networks*, Prentice-Hall, 1981.

3

Queueing systems: average performance

We shall be concerned with systems whose performance is affected by demands competing for the available servers. Jobs that cannot begin service immediately, or whose service is interrupted, have to wait in one or more queues. A scheduling policy controls the order in which waiting jobs are selected for service. These are called 'Queueing Systems', and their study is the subject of 'Queueing Theory'.

There is a commonly used shorthand notation for specifying queueing system models. In its simplest version it consists of three parts, of the form $\cdot / \cdot / \cdot$. The first part describes the nature of the arrival process, the second describes the distribution of service times, and the third indicates the number of servers. Letters such as M, D or G are used as abbreviations for 'Memoryless', 'Deterministic' or 'General', respectively. Thus, an M/D/2 system is one where the arrival process is memoryless (i.e. Poisson), the service times are deterministic (i.e. constant) and there are two parallel servers; in a G/M/1 system, the interarrival intervals have a general distribution, the service times are memoryless (i.e. exponentially distributed), and there is a single server. Sometimes further parts are added to the notation in order to specify particular features (e.g. a finite waiting room or a finite number of users submitting jobs).

Most of the models examined in this chapter are of type M/G/1. In some of them there is more than one Poisson process of arrivals. Different scheduling policies are analysed and evaluated, and some optimization problems are addressed. The performance measures of interest are various long term averages: average number(s) of jobs waiting and/or receiving service, average delays in different parts of the system, etc.

The principal tool used in the derivations, apart from the properties of the Poisson process, is a fundamental relation between three important

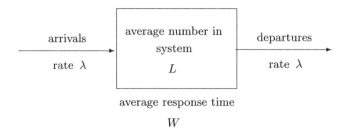

Fig. 3.1. A system with arrivals and departures

characteristics of an arbitrary queuing system. That relation, together with some general conclusions that can be drawn from it, is presented first.

3.1 Little's theorem and applications

Consider a system where jobs arrive, remain for some time, and then depart. The internal structure is not important; what happens to each job in the interval between its arrival and departure is immaterial; the arrival process need not be renewal. The only assumption we make is that the system is in steady state, or statistical equilibrium.

Steady state. Intuitively, this means that the system has been running for a long time and its behaviour no longer exhibits any trends. It continues to change state—the number of jobs in it continues to vary—but the probability of observing it in any given state is no longer a function of time (a formal definition will be given in chapter 5). Systems which are able to reach steady state are said to be 'stable'.

The above abstract system is illustrated in figure 3.1. The following aspects of steady-state behaviour should be emphasized:

1. The arrival rate, λ, does not change with time and is equal to the departure rate. In some contexts that quantity is called the 'system throughput'.
2. The average number of jobs in the system, L, does not change.
3. The average interval, W, between the arrival of a job and its departure (the 'response time') does not change.

In some models, λ is a given parameter, while L and W are performance measures to be determined by analysis. In others, all three quantities are sought. In all cases, the following relation holds:

Theorem 3.1 (J.D.C. Little) $L = \lambda W$.

This result existed as a 'folk theorem' before it was first proved by Little in 1961. Since then, several other proofs have been proposed, each of them simpler and more elegant than the preceding. Here we offer two very elementary arguments: one analytic and one economic.

Denote by $F(x)$ the probability distribution function of the response time. Let t be an arbitrary point in the steady state. The jobs that are in the system at time t are those that arrive before t and depart after t. Consider an infinitesimal interval at distance x in the past, $(t - x, t - x + dx)$. An average of λdx jobs arrive then, and each of them is still in the system at time t with probability $1 - F(x)$. Hence, that interval contributes $\lambda[1 - F(x)]dx$ to the average L. Integrating over all x from 0 to ∞, we get

$$L = \int_0^\infty \lambda[1 - F(x)]dx = \lambda W . \qquad (3.1)$$

The economic argument is even more straightforward. Suppose that a charge of £1 is levied on each job in the system, for each unit of time that it spends there. Then, if the money is collected continuously, the total average revenue per unit time is equal to the average number of jobs present, which is L. On the other hand, the entire amount charged on one job is equal to the time it spends in the system, which is W on the average. That entire amount can be collected when the job arrives (or departs). Since an average of λ jobs arrive (and leave) per unit time, this alternative collection method yields an average revenue of λW per unit time. Clearly, in the long run, the revenues produced by the two methods are the same. Hence $L = \lambda W$. qed.

Little's theorem is very general. It involves only averages and the assumption of stationarity. The internal structure of the system is immaterial: it may contain any number of servers and queues, with arbitrary interconnections among them; any scheduling policy may be employed, and there may be dependencies between jobs. The interarrival and service times may have general distributions. In fact, we are free to define both the system and the jobs in it as we wish, provided that the quantities L, λ and W are defined consistently. That freedom will be exploited to considerable advantage.

3.1 Little's theorem and applications

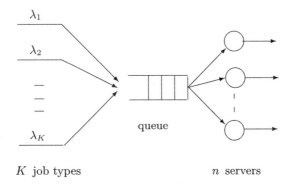

Fig. 3.2. A queueing system with K types of jobs and n servers

3.1.1 Utilization and response time laws

A basic performance measure of any queueing system is the steady-state utilization, U, of its servers. When there is only one server, U is defined as the probability that it is busy. If service is provided by n identical parallel servers, r of which are busy on the average, then $U = r/n$ (the two definitions coincide for $n = 1$). More generally, suppose that the demand contains jobs of K different types and service is provided by n identical parallel servers, as illustrated in figure 3.2. Denote by r_i the average number of servers busy with jobs of type i ($i = 1, 2, \ldots, K$). Then the server utilization due to type i is defined as $U_i = r_i/n$; the total utilization is $U = U_1 + U_2 + \ldots + U_K$. These definitions imply that $U \leq 1$.

Let λ_i and b_i be the arrival rate and the average service requirement for jobs of type i, respectively. No assumptions are made about the distributions of interarrival intervals or services. Each server can serve one job at a time. Each job in the system is either in the queue (i.e. waiting for service), or being served. The scheduling policy can be arbitrary and may involve service interruptions, so long as the latter are not attended by loss of service. Servers cannot be idle if there are jobs waiting; jobs do not depart before they are completed.

The product, $\rho_i = \lambda_i b_i$, represents the total average demand for service of type i that arrives into the system per unit time (seconds of service per second). That quantity is called the 'offered load' of type i; the sum $\rho = \rho_1 + \rho_2 + \ldots + \rho_K$ is the 'total offered load'.

Denote by W_i and w_i the average amounts of time that type i jobs spend in the system, and waiting for service, respectively. The above assumptions imply that

$$W_i = w_i + b_i \;;\; i = 1, 2, \ldots, K, \tag{3.2}$$

even if the scheduling policy obliges some jobs to alternate more than once between periods of waiting and service. Similarly, denote by L_i and l_i the average numbers of type i jobs in the system, and waiting for service, respectively. Since r_i is the average number of servers busy with type i jobs, we have

$$L_i = l_i + r_i \;;\; i = 1, 2, \ldots, K. \tag{3.3}$$

Apply Little's theorem to the following two 'systems': the first consists of the type i jobs waiting for and/or receiving service; the second contains only the type i jobs that are waiting for service. This yields

$$L_i = \lambda_i W_i \;;\; i = 1, 2, \ldots, K, \tag{3.4}$$

$$l_i = \lambda_i w_i \;;\; i = 1, 2, \ldots, K. \tag{3.5}$$

Subtracting (3.5) from (3.4) and using (3.3) and (3.2), we obtain

$$r_i = \lambda_i b_i = \rho_i \;;\; i = 1, 2, \ldots, K. \tag{3.6}$$

This result is known as the

Utilization law. In a general queueing system with K job types and n parallel servers, the average number of servers busy with type i jobs is equal to the offered load of type i. The utilization due to type i is $U_i = \rho_i/n$ and the total utilization is equal to $U = \rho/n$.

An obvious consequence of the utilization law is that if steady state exists, the total offered load does not exceed the number of servers: $\rho \leq n$. In other words, this inequality is a necessary condition for the stability of the system.

A single server. In the special case where $n = 1$, U_i is the probability that a job of type i is being served. According to the utilization law, that probability is equal to the offered load of type i, ρ_i.

Infinitely many servers. When $n = \infty$, there is no queue. All jobs in the system receive service in parallel, regardless of their number. The average response time for type i jobs is equal to their average service time: $W_i = b_i$. The utilization law implies that the average number of

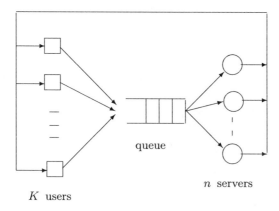

Fig. 3.3. A finite-source system

type i jobs in the system is equal to the corresponding offered load: $L_i = r_i = \rho_i$. The system is always stable.

A 'queueing' system with infinitely many servers will in future be called an 'independent delay' system. This is to emphasize the fact that jobs do not compete for service, but are delayed for random intervals, independently of each other.

Now suppose that, instead of arriving in Poisson processes with fixed rates, jobs are generated by a finite number, K, of identically behaved users: each user, having submitted a job, waits until it is completed, then 'thinks' for an interval of time, then submits another job, etc. Service is provided by n identical parallel servers (see figure 3.3).

This may be a model of a transaction system, where remote users send instructions to a central multiprocessor computer. Alternatively, it could be a maintenance model: K machines break down from time to time and are attended by n repairmen (the think times are machine operative periods and the jobs are repair tasks).

The average think intervals, τ, and the average required service times, b, are parameters in this system. The performance measures are the utilization of the servers, U, the average response time, W (this is the interval between submitting a job and its completion), and the throughput, T (the average number of job completions per unit time). None of those performance measures can be determined without additional assumptions. However, there are useful relationships among them that have general validity.

80 3 Queueing systems: average performance

Applying Little's theorem to the subsystem consisting of the servers and the queue, we can write

$$L = TW, \quad (3.7)$$

where L is the average number of jobs waiting for or receiving service. The arrival rate into the subsystem is T, because in the steady state the average number of completed jobs per unit time and the average number of submitted jobs per unit time are equal.

Similarly, applying Little's theorem to the subsystem consisting of the thinking users (their average number is $K-L$ and the average think time is τ), gives

$$K - L = T\tau. \quad (3.8)$$

Eliminating L from (3.7) and (3.8) yields the following relation between the average response time and the throughput:

Response time law.

$$W = \frac{K}{T} - \tau. \quad (3.9)$$

A different form of the response time law is obtained by remarking that, since the average service time is b, each server completes jobs at rate $1/b$ while it is busy. If the utilization of the servers is U, the average number of busy servers is nU and so the throughput is equal to $T = nU/b$. Hence,

$$W = \frac{bK}{nU} - \tau. \quad (3.10)$$

Consider the behaviour of the average response time as K increases, other parameters remaining fixed. It is clear that the more users submit jobs, the busier the servers become: $U \to 1$ when $K \to \infty$. The response time law (3.10) indicates that W becomes approximately a linear function of K, with slope b/n.

Exercises

1. The demand to see an exhibit in a museum is so heavy that the management decides to operate an admission policy which restricts the number of people in that room to 60. When a visitor leaves, a new one is admitted from among those waiting outside. The resulting throughput turns out to be 5 visitors per minute. How long does each visitor remain in the room on the average?

Fig. 3.4. The M/G/1 queue

2. (Example 2 in section 2.3.1 revisited). Machine parts arrive on a conveyor belt to a robot operator at arbitrarily distributed intervals with mean 50 seconds. The operator works on each part for an arbitrarily distributed interval with mean 30 seconds. What is the long term probability that a random observer will see a busy operator, and why?

3. An airline reservation system handles transactions from 100 terminals on an 8-processor computer. The average transaction run times and user think times are 2 minutes and 15 minutes, respectively. In an effort to improve performance, the company decides to upgrade to a 16-processor computer; this reduces the processor utilization by 25% and the average response time by 10 minutes. What were the values of U and W before the upgrade, and what are they after?

3.2 The single-class M/G/1 queue

We have seen that an arrival process formed by merging together requests from a large number of independent sources is approximately Poisson (see theorem 2.4 in section 2.3.1). There are thus many systems where jobs can reasonably be assumed to arrive according to a Poisson process with some rate, λ. Assume further that there is a single server, and that service times are independent and identically distributed random variables with some general distribution function, $F(x)$. Jobs wait in an unbounded queue and are served without interruptions, in the order in which they arrive; this scheduling policy is referred to as FIFO (First-In-First-Out), or FCFS (First-Come-First-Served).

The above model is known as the M/G/1 queue (figure 3.4). The term 'single-class' in the section heading refers to the fact that all jobs are statistically identical and are treated equally.

Denote, as before, the average required service time by b. The offered load (the average amount of work arriving into the system per unit time) is $\rho = \lambda b$. According to the utilization law, that is also equal to the average number of jobs in service, or the probability that the server is busy. The stability condition for this system is $\rho < 1$. We shall assume that that condition is satisfied, and that the queue is in steady state.

The performance measures of interest are:

L = average number of jobs in the system;
l = average number of jobs waiting for service;
W = average response time (interval between the arrival of a job and its departure);
w = average waiting time (interval between the arrival of a job and its starting service).

Of course, these quantities are related:

$$W = w + b \; ; \; L = l + \rho , \qquad (3.11)$$

by the definitions and the utilization law. Also,

$$L = \lambda W \; ; \; l = \lambda w , \qquad (3.12)$$

by Little's theorem.

Now let us 'tag' a job arriving into the system and consider the average delay that it may experience before it can start service. It is convenient to break this into two components: (i) the average delay, w_0, caused by the job (if any) whose service is in progress when the tagged job arrives; (ii) the average delay, A, due to the jobs found waiting. Thus,

$$w = w_0 + A , \qquad (3.13)$$

According to the PASTA property of the Poisson process (section 2.3.1), the tagged job can be treated like a random observer of the system. It sees a job in service with probability ρ. If it does see a service in progress, the latter's remaining duration—its residual lifetime—is equal to $M_2/(2b)$, where M_2 is the second moment of the service time distribution (section 2.1.1). Hence,

$$w_0 = \rho \frac{M_2}{2b} = \frac{\lambda M_2}{2} . \qquad (3.14)$$

The tagged job sees an average of l jobs waiting in the queue. Since each of those jobs takes an average time b to serve, the second component of the delay is $A = lb$. Substituting the last two expressions into (3.13), and using (3.12) to replace l with λw, we get

$$w = \frac{\lambda M_2}{2} + \lambda w b = \frac{\lambda M_2}{2} + \rho w . \qquad (3.15)$$

An unknown quantity, w, has thus been expressed in terms of itself.

3.2 The single-class M/G/1 queue

This is a standard technique, which we shall have occasion to use again. Relation (3.15) is a simple example of a so-called 'fixed-point' equation.

Solving (3.15) for w, together with (3.11) and (3.12), yields all performance measures:

$$w = \frac{\lambda M_2}{2(1-\rho)} \; ; \; W = \frac{\lambda M_2}{2(1-\rho)} + b \; . \tag{3.16}$$

$$l = \frac{\lambda^2 M_2}{2(1-\rho)} \; ; \; L = \frac{\lambda^2 M_2}{2(1-\rho)} + \rho \; . \tag{3.17}$$

These results are known as the 'Pollaczek–Khinchin formulae'. They are often presented in a slightly different form, using the squared coefficient of variation of the service time, $C^2 = (M_2 - b^2)/b^2$ (section 1.3.2), instead of the second moment. After substitution of $M_2 = b^2(1+C^2)$, (3.16) and (3.17) become

$$w = \frac{b\rho(1+C^2)}{2(1-\rho)} \; ; \; W = \frac{b\rho(1+C^2)}{2(1-\rho)} + b \; . \tag{3.18}$$

$$l = \frac{\rho^2(1+C^2)}{2(1-\rho)} \; ; \; L = \frac{\rho^2(1+C^2)}{2(1-\rho)} + \rho \; . \tag{3.19}$$

Note that under the FIFO scheduling policy, the waiting time of a job does not depend on its required service time. Denote the conditional average response time of a job, given that it requires service x, by $W(x)$. That performance measure has a simple expression:

$$W(x) = w + x = \frac{\lambda M_2}{2(1-\rho)} + x \; . \tag{3.20}$$

So, the parameters that influence the average performance of the M/G/1 system are the arrival rate, λ, the average service time, b, and the second moment, M_2 (or the squared coefficient of variation, C^2), of the service time. When the offered load approaches 1, both the average delays and the average queue size grow without bound. In a heavily loaded system, a small increase in traffic can have a disastrous effect on performance.

The shape of the service time distribution function affects performance only through the second moment; higher moments do not matter. Even a lightly loaded M/G/1 queue can perform very poorly when the variability of the demand is high. For fixed values of λ and b, the best performance is achieved when $C^2 = 0$, i.e. when there is no variability in the service times: all jobs require exactly the same amount of service.

84   3 Queueing systems: average performance

Examples

1. A bad post office. Customers arrive at a single post office counter in a Poisson process with rate 1 customer per minute. The service times have two possible values: 99% of them are 20 seconds long and 1% are 30 minutes. Thus, the average service time is

$$b = 0.99 \times 20 + 0.01 \times 1800 = 19.8 + 18 = 37.8 \text{ sec}.$$

Here $\lambda = 1/60$ and $\rho = 37.8/60 = 0.63$. This is not a particularly heavily loaded system. To find the average response time, we need the second moment of the service time:

$$M_2 = 0.99(20)^2 + 0.01(1800)^2 = 396 + 32400 = 32796.$$

Applying the Pollaczek–Khinchin formula we find the average time customers spend in the post office:

$$W = \frac{32796}{2(1 - 0.63)60} + 37.8 \approx 776.5 \text{ sec} \approx 12.9 \text{ min}.$$

The average number of customers present is also 12.9.

2. A good post office. An almost identical branch at the other end of town has the same arrival rate of customers and the same average service time, but a different distribution: here all service times are equal to 37.8 seconds. Now $M_2 \approx 1428.8$, and the average response time is

$$W \approx \frac{1428.8}{2(1 - 0.63)60} + 37.8 \approx 70 \text{ sec} \approx 1.2 \text{ min}.$$

This is the best achievable performance for the given values of λ and b.

3. A FIFO disk storage device. On the surface of a computer disk there is a large number of concentric circular tracks, each of which is capable of storing digital information. An input/output request is addressed to a particular record on a particular track. To perform a read or a write operation, the read/write head has to move forward or backward until it is positioned over the desired track; this is called the 'seek delay'. Then there is a 'rotational delay', until the start of the desired record comes under the head. Finally, the record is read or written as it passes by (see figure 3.5).

Assume that jobs (input/output requests) arrive in a Poisson process with rate λ and are served one at a time in FIFO order. Each service

3.2 The single-class M/G/1 queue

seek delay   rotational delay

Fig. 3.5. Disk storage device with moving read/write head

time, S, is the sum of three random variables:

$$S = X_1 + X_2 + X_3 ,$$

where X_1 is the seek delay, X_2 is the rotational delay and X_3 is the read/write time. Assume, for simplicity, that all records are of equal length and that there are n of them on each track. Then, if the rotational speed of the disk is R revolutions per second, we have $X_3 = 1/(nR)$.

Assume that when the head reaches the target track, the disk is equally likely to be in any rotational position. Then the rotational delay is equally likely to take any value between 0 and one revolution, i.e. X_2 is uniformly distributed on the interval $(0, 1/R)$. Its mean and second moment are

$$E(X_2) = \frac{1}{2R} \; ; \; E(X_2^2) = \frac{1}{3R^2} .$$

The simplest way to model the seek delay is to treat the track positions of the head as points on the interval $(0, 1)$. If the tracks addressed by consecutive requests are assumed to be independent of each other and uniformly distributed, the distance travelled by the head on each seek is approximately the distance between two independent random points uniformly distributed on the interval $(0, 1)$. Assuming further that the head travels at constant speed such that it moves from the innermost to the outermost track (or vice versa) in time d, we can say that X_1 is the distance between two independent random points uniformly distributed on the interval $(0, d)$. It is not difficult to establish that the mean and second moment of X_1 are

$$E(X_1) = \frac{d}{3} \; ; \; E(X_1^2) = \frac{d^2}{6} .$$

Now we have the average service time:

$$b = E(S) = \frac{d}{3} + \frac{1}{2R} + \frac{1}{nR} .$$

The condition for stability is that the arrival rate should satisfy the inequality $\lambda < 1/b$. The second moment of the service time is given by

$$M_2 = E(S^2) = E(X_1^2) + E(X_2^2) + E(X_3^2)$$
$$+ 2E(X_1)E(X_2) + 2E(X_1)E(X_3) + 2E(X_2)E(X_3) \ .$$

When the revolution time is small compared to the seek time, the terms involving $1/R^2$ can be neglected. This yields,

$$M_2 \approx \frac{d^2}{6} + \frac{d}{3R} + \frac{2d}{3nR} \ .$$

The above values for b and M_2 can be substituted into (3.16) and (3.17) to determine the performance measures.

It should be pointed out that the assumption of uniform and independent accesses across tracks is a rather crude simplification; more realistic, but also more complicated, analyses of disk performance have been carried out. Also, the FIFO policy is by no means the best way of scheduling requests; there are various alternatives which lead to fewer and shorter head movements.

3.2.1 Busy periods

The server of the M/G/1 queue goes through alternating periods of being idle and busy. An idle period starts with the departure of a job leaving an empty queue, and ends with the arrival of the next job; because of the memoryless property of the exponential distribution, the average length of an idle period is $1/\lambda$.

A busy period starts with the arrival of a job which finds an empty queue, and ends with the departure of the next job which leaves an empty queue. Denote its average length by B. More generally, an interval between an instant when there are j jobs present in the system and one of them is starting service, and the next departure which leaves an empty queue, is called a 'busy period of order j' (figure 3.6). let B_j be the latter's average length. Obviously, a busy period is a busy period of order 1.

Remark. An important property of any busy period of any order is that it does not depend on the scheduling policy. As long as the server is obliged to work whenever there are jobs present, and jobs are not allowed to leave the system before they are completed, the busy period is determined by the instants of arrival and the required service times. Exactly which job is being served at any moment is not important. The

3.2 The single-class M/G/1 queue

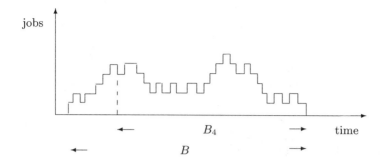

Fig. 3.6. Busy period of order 4 within a busy period

policy may also allow interruptions of service, provided that the latter are eventually resumed without loss.

Consider the time, $X_{j,j-1}$, taken to reduce the number of jobs in the system from j to $j-1$. This is an interval which starts with a job beginning its service and $j-1$ jobs waiting in the queue, and ends with the next departure that leaves $j-1$ jobs in the system. Because of the independence remarked above, we can assume that subsequent arrivals (if any) during such an interval are served before the original $j-1$ jobs in the queue. Then we see that $X_{j,j-1}$ is independent of j and is equal, in distribution, to a busy period. Hence, a busy period of order j can be regarded as the sum of j independent busy periods. The averages satisfy

$$B_j = jB \ . \tag{3.21}$$

Now, a busy period starts with the arrival of a job into an empty system. If, by the end of that job's service, j further jobs have arrived, then the rest of the busy period will be a busy period of order j, and the average length of that remainder will be jB. Since the average service time is b, and during it an average of λb jobs arrive (see section 2.3.2), we can write

$$B = b + \lambda bB \ . \tag{3.22}$$

(A simpler way of arriving at the same equation would be to argue that the busy period consists of the service of the initial job, plus the services of all jobs that arrive during the busy period; the average number of those arrivals is λB. However, the above argument can be generalized to derive the distribution and higher moments of the busy period; see exercise 4.)

The solution of (3.22), together with (3.21), gives

$$B = \frac{b}{1-\rho} \; ; \; B_j = \frac{jb}{1-\rho} \; ; \; j = 2, 3, \ldots , \qquad (3.23)$$

where $\rho = \lambda b$ is the offered load. The average number of jobs served during a busy period, n_B, is obtained from the fact that $n_B b = B$:

$$n_B = \frac{1}{1-\rho}. \qquad (3.24)$$

Remark. It is notable that B and B_j do not depend on the distribution of service times; only the mean, b, appears in the expressions. The average number of jobs served during a busy period does not even depend on the individual values of b and λ; it is determined by the offered load.

3.2.2 The M/M/1 queue

The performance measures given by the Pollaczek–Khinchin expressions have a particularly simple form when the required service times are distributed exponentially. The special case of the M/M/1 queue is of interest in its own right, because the exponential assumption is the appropriate one to adopt in many cases, and also because this was the first queueing model to be analysed mathematically.

The second moment and the squared coefficient of variation of the exponential distribution are $M_2 = 2b^2$ (section 2.2), and $C^2 = 1$. Substituting the latter into (3.18) and (3.19), we obtain

$$w = \frac{b\rho}{1-\rho} \; ; \; W = \frac{b}{1-\rho}. \qquad (3.25)$$

$$l = \frac{\rho^2}{1-\rho} \; ; \; L = \frac{\rho}{1-\rho}. \qquad (3.26)$$

Comparing (3.25) with (3.23), we note that the average response time and the average busy period in an M/M/1 system are equal. There does not seem to be an intuitive explanation of this coincidence. The corresponding distributions are not the same.

For an M/M/1 queue, it possible to find distributions of performance measures, as well as averages, by using very simple tools. Imagine that the jobs in the system occupy numbered positions, $1, 2, \ldots$: 1 is the service position, 2 is the first waiting position, etc. If there are n jobs present, positions $1, 2, \ldots, n$ are occupied and positions $n+1, n+2, \ldots$ are free. The FIFO policy implies that (a) an arriving job joins the first free position; (b) when a service is completed, if position j is occupied, the job in it moves to position $j-1$ ($j > 1$); the job in position 1 departs.

Denote by p_n the steady-state probability that there are n jobs in the system, and by q_n the probability that position n is occupied. Clearly,

$$q_n = \sum_{j=n}^{\infty} p_j \ ; \ p_n = q_n - q_{n+1} \ . \tag{3.27}$$

Apply Little's theorem to the subsystem consisting of position n ($n = 1, 2, \ldots$), in the steady state. The average number of jobs in that subsystem is equal to the probability that position n is occupied, which is q_n. To find the arrival rate, note that every job which finds, on arrival, $n-1$ or more jobs present, joins position n eventually. Therefore, according to the PASTA property of the Poisson process, the arrival rate into position n is equal to λq_{n-1}, where $q_0 = 1$ by definition. By the memoryless property of the exponential distribution, the average time that a job spends in position n is equal to b, regardless of how it got there. Hence,

$$q_n = \lambda q_{n-1} b = \rho q_{n-1} \ ; \ n = 1, 2, \ldots \ . \tag{3.28}$$

These recurrences, together with $q_0 = 1$ and (3.27), imply that

$$q_n = \rho^n \ ; \ p_n = \rho^n - \rho^{n+1} = \rho^n(1-\rho) \ ; \ n = 0, 1, \ldots \ . \tag{3.29}$$

So, the number of jobs in the system has the modified geometric distribution (example 3 in section 1.4.1). This has implications for the distribution of the response time. The interval between the arrival of a job and its departure is distributed like the sum of $N+1$ service times, where N is the number of jobs present on arrival. Since the distribution of N is modified geometric, that of $N+1$ is geometric, with parameter $1-\rho$:

$$P(N+1 = n) = \rho^{n-1}(1-\rho) \ ; \ n = 1, 2, \ldots \ .$$

The sum of a geometrically distributed number (parameter $1-\rho$) of independent and exponentially distributed random variables (parameter $1/b$) is distributed exponentially with parameter $(1-\rho)/b$ (exercise 5 in section 2.2). Thus, the distribution function, $G(x)$, of the response time is

$$G(x) = 1 - e^{-(1-\rho)x/b} \ ; \ x \geq 0 \ . \tag{3.30}$$

3 Queueing systems: average performance

Example

4. A central distribution depot receives deliveries by lorry from a number of sources. Assume that lorries from each source arrive in an independent Poisson process, with an average interarrival interval of 12 hours. There is a single unloading facility which works day and night, serving lorries in FIFO order, regardless of origin; the service times can be assumed exponentially distributed, with mean 0.5 hours.

Of interest is the largest number of sources that can be accepted by the depot. The following three criteria are suggested as possible grounds for reaching a decision.

1. The depot must not be overloaded.
2. The average response time for a lorry must be less than 2 hours.
3. Less than 5% of all lorries may have response times exceeding 2 hours.

To answer these questions, we model a system with K sources as an M/M/1 queue with parameters $\lambda = K/12$ and $b = 0.5$. This is valid because the superposition of independent Poisson processes is also a Poisson process (section 2.3.1).

The load generated by the K sources is $\rho = K/24$. Hence, in order to satisfy the first criterion, K must be less than 24.

The average response time requirement is expressed, using (3.25), as

$$\frac{0.5}{1 - (K/24)} = \frac{12}{24 - K} < 2 \ .$$

Solved for K, this yields $K < 18$.

If the third criterion is adopted, then the probability that the response time of a lorry is greater than 2 must be less than 0.05. In other words, the distribution function (3.30) must satisfy $1 - G(2) < 0.05$:

$$e^{-2(24-K)/12} < 0.05 \ .$$

The integer solution of the above inequality is $K \leq 6$.

These numbers illustrate the extent to which the third requirement is stronger than the second, which is itself stronger than the first.

Exercises

1. Two applicants are being considered for a vacancy as a bank teller. Tests reveal that the service times of the first applicant are distributed

3.2 The single-class M/G/1 queue

exponentially with mean 0.9 minutes, while those of the second are uniformly distributed between 0.8 and 1.2 minutes. It is known that customers arrive at that particular teller's window in a Poisson process, at the rate of 50 per hour.

Can both candidates cope with the load? If so, which one should be employed in order to minimize the average queue size?

2. Messages arrive into a single-server gateway in a Poisson stream, at the rate of 2 per second, and join an unbounded FIFO queue. The processing of each message consists of two consecutive phases: the first is constant, of duration 0.1 second, and the second is random, uniformly distributed between 0 and 0.6 seconds. Upon completion of the second phase, the message departs.

(a) Is this system stable, and what is the utilization of the server?

(b) Suppose that the system is observed at random. If a message is being processed, what is the average period until its departure?

(c) What is the steady-state average response time?

(d) If the arrival rate, λ, can be controlled, what condition should it satisfy so that the average response time does not exceed 1 second?

3. Consider a realization (a sample path) of a queuing process where jobs arrive and depart singly. Let $N(t)$ be the number of jobs present at time t. This is a step function which jumps up by 1 at arrival instants and down by 1 at departure instants. Assume that $N(0) = 0$ and that $N(t)$ keeps reaching the value 0 from time to time.

Show that to each step that $N(t)$ makes from j to $j+1$, there corresponds a step from $j+1$ to j, and vice versa ($j = 0, 1, \ldots$). Hence argue that, in the long run, the probability that an arriving job sees state j is equal to the probability that a departing job leaves state j behind.

4. Denote by $f^*(s)$ the Laplace transform of the M/G/1 service time density function, and by $\varphi(s)$ and $\varphi_j(s)$ the Laplace transforms of the pdfs of the busy period and the busy period of order j, respectively. First show that $\varphi_j(s) = \varphi(s)^j$. Then argue that a busy period consists of the initial service time and, if j jobs arrive during it, a busy period of order j. Condition upon the length of that first service and upon j, and demonstrate that $\varphi(s)$ satisfies the following functional equation:

$$\varphi(s) = f^*[s + \lambda - \lambda\varphi(s)].$$

Hence, by taking derivatives at $s = 0$, determine the second moment of the busy period in terms of λ, b and M_2.

5. Use the result from exercise 3 to prove that the interval between two consecutive departures from the M/M/1 queue is distributed exponentially with parameter λ. (Hint: denote by $\alpha(s)$, $\beta(s)$ and $\gamma(s)$ the Laplace transforms of the interarrival interval, service time and interdeparture interval, respectively. Show that

$$\gamma(s) = p_0 \alpha(s)\beta(s) + (1 - p_0)\beta(s),$$

where p_0 is the probability of an empty system. Hence establish that $\gamma(s) = \lambda/(\lambda + s)$.)

3.3 Different scheduling policies

There are applications of the M/G/1 model where the scheduling policy is not FIFO. In certain manufacturing processes, items requiring service are stored in such a way that the next item to be served is the one that arrived most recently. This scheduling policy is called Last-In-First-Out; in its context, the queue of waiting jobs is often referred to as the 'stack'. We shall consider two versions:

(a) Ordinary, or non-preemptive Last-In-First-Out (LIFO). Services are not interrupted. An incoming job that finds the server busy is placed on top of the stack.

(b) Preemptive-Resume Last-In-First-Out (LIFO-PR). An incoming job interrupts the current service, if any, and engages the server. Interrupted jobs are placed on top of the stack; eventually they resume their service from the point of interruption.

In both cases, whenever a service is completed, the server starts serving the job currently on top of the stack (figure 3.7).

The other assumptions are the same as before: the arrival process is Poisson with rate λ; service times are generally distributed, with mean b and second moment M_2.

Let us tag a job and find the delays it experiences under the non-preemptive LIFO policy. First, there may be a job in service at the time of arrival. The average delay associated with the residual service,

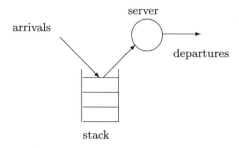

Fig. 3.7. The LIFO scheduling policy

denoted by w_0, is the same for all non-preemptive policies; it is given by (3.14). If no other jobs arrive during w_0, the tagged job will remain on top of the stack and will start its own service. Otherwise, if j jobs arrive, they will be served before the tagged job, as will all arrivals during *their* services, etc. In fact, the remaining waiting time of the tagged job will then be a busy period of order j, whose average length is given by (3.23).

Since the average number of arrivals during an interval of average length w_0 is λw_0, we can express the total average waiting time of the tagged job in terms of the average busy period and hence, using (3.14) and (3.23), in terms of the service time parameters and offered load:

$$w = w_0 + \lambda w_0 B = \frac{\lambda M_2}{2(1-\rho)}. \qquad (3.31)$$

This is the same expression as for the FIFO policy. Consequently, the average response time, W, the conditional average response time, $W(x)$, and average numbers of jobs in the queue, l, and in the system, L, are also the same under the two policies. However, there are no analogous equalities for the corresponding distributions.

Now consider the path of a tagged job under the LIFO-PR policy. As soon as it arrives, the job starts service, regardless of whether the server was busy. That service may be interrupted by jobs arriving while it is in progress. Each such interruption consists of the service of the interrupting job, plus the services of the jobs arriving during *that* service, plus the services of the jobs arriving during *those* services, etc. In other words, each interruption is equivalent to a busy period.

The average number of arrivals during the service of the tagged job is λb. Therefore, the total average time that the tagged job remains in the system is

$$W = b + \lambda b B = \frac{b}{1-\rho}. \qquad (3.32)$$

The average number of jobs in the system is given by

$$L = \lambda W = \frac{\rho}{1-\rho}. \qquad (3.33)$$

Note that these performance measures do not depend on the service time distribution, once the mean is fixed. The M/M/1, M/D/1 or any other M/G/1 queue has the the same average performance under the LIFO-PR policy. Scheduling policies of that type are said to be 'insensitive'.

The conditional average response time, $W(x)$, given that the job's required service time is x, is obtained in a similar way to (3.32):

$$W(x) = x + \lambda x B = \frac{x}{1-\rho}. \qquad (3.34)$$

3.3.1 The Processor-Sharing policy

In computer systems where a single processor provides service to a number of jobs running concurrently, the following scheduling policy is usually employed. The processor offers service in quanta of fixed size, Q. The job at the head of the queue is given a quantum of service. If that job completes before the quantum elapses, it departs; otherwise it returns to the end of the queue and awaits its turn again. Thus, if there are n jobs competing for the processor, each of them occupies it for one out of every n quanta. This policy is called 'Round-Robin'.

The Round-Robin policy can be analysed, under suitable assumptions, by following the progress of a tagged job as it circulates round the system. Such an analysis tends to be rather involved, because the successive passes that the job makes through the queue are not independent of each other. Much of the complexity disappears, however, if the quantum size Q is allowed to shrink to 0. The smaller the quantum, the more frequently each waiting job visits the processor. In the limit $Q \to 0$, the picture of frantic circulation blurs into one where all competing jobs run smoothly in parallel, but at a slower pace than if they were on their own. That policy is called Processor-Sharing (PS). It is a good approximation of Round-Robin with a small quantum, and is easier to analyse (figure 3.8).

3.3 Different scheduling policies

Fig. 3.8. The Round-Robin and Processor-Sharing policies

The PS policy can be defined directly by saying that the available processing capacity is divided equally among the jobs present. If there are n jobs in the system throughout an interval of length x, then during that interval each of the n jobs receives an amount of service equal to x/n (i.e. each job receives service at rate $1/n$). Alternatively, the time necessary for a job to receive service x, given that it is sharing the server with $n-1$ other jobs, is equal to nx.

Jobs arrive in a Poisson process with rate λ, and their required service times are generally distributed, with mean b. The performance measures of interest are again the average response time, W, the average number of jobs present, L, and the conditional average response time, $W(x)$, given that the required service time is x. We shall derive them by means of an intuitive argument.

Follow the progress of a tagged job whose service requirement is x. It finds, on arrival, an average of L jobs already present (PASTA holds regardless of x). Since the system is in steady state, the tagged job shares the processor with L other jobs, on the average, until it is completed. On the other hand, the time taken by a job to attain service x when $L+1$ jobs share the processor is $(L+1)x$. Hence,

$$W(x) = (L+1)x = (\lambda W + 1)x \,. \tag{3.35}$$

Multiplying both sides of (3.35) by the service time pdf and integrating over all values of x, we get an equation for the unconditional average response time, W:

$$W = (\lambda W + 1)b \,. \tag{3.36}$$

This yields an expression which is by now familiar:

$$W = \frac{b}{1-\rho} \,. \tag{3.37}$$

Going back to (3.35), $W(x)$ becomes

$$W(x) = \frac{x}{1-\rho} \,. \tag{3.38}$$

96   3 Queueing systems: average performance

The weak point in the above argument is the assertion (not quite an obvious one) that the average number of jobs with which the tagged job shares the processor remains constant throughout the latter's residence in the system. However, both that assertion and the resulting expressions are correct and can be established rigorously.

It is notable that the PS policy, like the LIFO-PR one, is insensitive to the service time distribution. There is in fact a whole class of policies, known as 'symmetric', which have similar properties.

3.3.2 Symmetric policies and multiclass queues

To define the class of symmetric scheduling policies, consider again the sequence of numbered positions which jobs may occupy while they are in the system. An incoming job which finds n jobs present on arrival, and therefore positions $1, 2, \ldots, n$ occupied, is allowed to join position j ($j = 1, 2, \ldots, n+1$) with probability $\alpha(j, n+1)$. Except for $j = n+1$, such an entry causes the jobs previously in positions $j, j+1, \ldots, n$ to move into positions $j+1, j+2, \ldots, n+1$, respectively.

Jobs in different positions may be served in parallel, perhaps at different rates. If n positions are occupied, the job in position j receives service at rate $\beta(j, n)$ ($j = 1, 2, \ldots, n$). If a job in position j departs, those in positions $j+1, j+2, \ldots$ move to positions $j, j+1, \ldots$, respectively.

A policy of this type is said to be symmetric if the probability of joining position j is equal to the service rate in position j:

$$\alpha(j, n) = \beta(j, n) \ ; \ j = 1, 2, \ldots \ ; \ n = j, j+1, \ldots \ . \tag{3.39}$$

The PS policy of the previous subsection is symmetric. It corresponds to the special case where a job may join any available position with equal probability, and all jobs in the system are served at equal rate:

$$\alpha(j, n) = \beta(j, n) = \frac{1}{n} \ .$$

The LIFO-PR policy is also symmetric. Arriving jobs always join position 1, and only the job in position 1 receives service:

$$\alpha(1, n) = \beta(1, n) = 1 \ ; \ \alpha(j, n) = \beta(j, n) = 0 \ ; \ j > 1 \ .$$

The FIFO policy is not symmetric, because position 1 is served at rate 1 ($\beta(1, n) = 1$), while new arrivals join the first free position ($\alpha(n, n) = 1$). The non-preemptive LIFO policy is not symmetric, because position 1 is served at rate 1, while incoming jobs join position 2.

3.3 Different scheduling policies

Any symmetric policy has the property that new arrivals start receiving service immediately, although they may have to share the server with other jobs. Policies which allocate service capacity on the basis of the types of jobs present (e.g. by means of priorities) cannot be symmetric.

We shall state the following results without proof. All symmetric policies are insensitive to the service time distribution. Their average performance depends only on the arrival rate, λ, and average service time, b:

$$L = \frac{\rho}{1-\rho} \ ; \ W = \frac{b}{1-\rho} \ ; \ W(x) = \frac{x}{1-\rho}. \quad (3.40)$$

The probabilities, p_n, that there are n jobs in an M/G/1 system under a symmetric scheduling policy are modified geometric:

$$p_n = \rho^n(1-\rho) \ ; \ n = 0, 1, \ldots. \quad (3.41)$$

More generally, suppose that the input to the queue consists of K job types arriving in Poisson processes with rates $\lambda_1, \lambda_2, \ldots, \lambda_K$, and requiring average service times $b_1, b_2, \ldots, b_K$, respectively. The offered load of type i is $\rho_i = \lambda_i b_i$. Then the joint probability that there are n_1 jobs of type 1, n_2 jobs of type 2, ..., n_K jobs of type K in the system is given by

$$p(n_1, n_2, \ldots, n_K) = (1-\rho)n! \prod_{i=1}^{K} \frac{\rho_i^{n_i}}{n_i!}, \quad (3.42)$$

where $\rho = \rho_1 + \rho_2 + \ldots + \rho_K$ and $n = n_1 + n_2 + \ldots + n_K$. This last result indicates that an arbitrary job observed in the system is of type i with probability ρ_i/ρ. See exercise 3 for the marginal distribution of the type i jobs.

Independent delay systems. A system with infinitely many servers (see section 3.1.1) is also symmetric. The 'scheduling policy' in this case simply gives every job a server, as soon as it arrives. This can be thought of as a form of Processor-Sharing, with $\alpha(j, n) = \beta(j, n) = 1/n$, except that the available processing capacity is equal to n when there are n jobs present. The joint distribution corresponding to (3.42) is now

$$p(n_1, n_2, \ldots, n_K) = \prod_{i=1}^{K} \frac{\rho_i^{n_i}}{n_i!} e^{-\rho_i}. \quad (3.43)$$

Not surprisingly, the number of jobs of type i in an independent delay system is independent of how many other jobs are present. The fact that the distribution of that number is Poisson with parameter ρ_i is less obvious. Because of the insensitivity property, it is enough to prove the result for one particular service time distribution. We shall do that for exponentially distributed service times in chapter 5.

3.3.3 Which policy is better?

It is worth comparing the performance of the M/G/1 queue under the FIFO or non-preemptive LIFO policy on one hand, and a symmetric policy like PS or LIFO-PR on the other. Take the average response time as the performance measure of interest, and ignore other factors (such as fairness or ease of implementation of the policies). Then the two expressions to be compared are, say, (3.18) and (3.37):

$$\frac{b\rho(1+C^2)}{2(1-\rho)} + b \quad \text{vs} \quad \frac{b}{1-\rho}. \qquad (3.44)$$

The outcome depends on the value of the squared coefficient of variation: FIFO and the non-preemptive LIFO are better if $C^2 < 1$, the symmetric policies are better if $C^2 > 1$.

Alternatively, the comparison may be based on the conditional average response time, $W(x)$. Now the two expressions are

$$\frac{b\rho(1+C^2)}{2(1-\rho)} + x \quad \text{vs} \quad \frac{x}{1-\rho}. \qquad (3.45)$$

Clearly, short jobs have lower average response times under PS and LIFO-PR, while long jobs are treated better under FIFO and the non-preemptive LIFO. The exact position of the boundary depends on the value of C^2. For instance, if the service times are distributed exponentially ($C^2 = 1$), then the jobs with shorter than average service times ($x < b$), and only they, are better off under the symmetric policies.

Exercises

1. The arrival rate in an M/G/1 queue is 6 jobs per hour. Two scheduling policies can be implemented, FIFO or LIFO-PR. Evaluate their performance and determine which is preferable, in each of the following cases:

(a) Service times are constant, equal to 8 minutes. The performance measure is the average response time.

(b) Service times may take two values, 2 and 14 minutes, with equal probability. The performance measure is the average response time of the 2-minute jobs.

(c) As (b), except that the performance measure is the average response time of the 14-minute jobs.

(d) Service times are distributed uniformly on the interval (0,16) minutes. The performance measure is the average response time.

2. From the definition of the conditional average response time it follows that $dW(x) = W(x + dx) - W(x)$ is the average time needed for a job whose required service time is greater than x, to increase its attained service from x to $x + dx$. Let $n(x)$ be the average number of jobs in the system whose attained service is less than or equal to x. Then $dn(x) = n(x + dx) - n(x)$ is the average number of jobs in the system whose attained service is in the interval $(x, x + dx)$. Apply Little's theorem to those jobs and show that

$$dn(x) = \lambda[1 - F(x)]dW(x),$$

where $F(x)$ is the distribution function of the required service times. Hence derive an expression for $n(x)$ in the case of an insensitive policy.

3. Consider the M/G/1 queue with K job types introduced in section 3.3.2. By summing the joint probabilities (3.42) over all vectors $n_1, n_2, \ldots, n_K$ such that $n_i = n$, for some i, show that the marginal probability that there are n jobs of type i in the system, $p_i(n)$, is given by

$$p_i(n) = \beta_i^n(1 - \beta_i) \; ; \; n = 0, 1, \ldots,$$

where $\beta_i = \rho_i/(1 - \rho + \rho_i)$. Note that, although the joint distribution (3.42) has the form of a product, it is not the product of the marginal distributions. The numbers of jobs of different types in the system are not independent of each other.

3.4 Priority scheduling

In many queueing systems of practical interest, the demand consists of jobs of different types. These job types may or may not have different arrival and service characteristics. Rather than treat them all equally and serve them in FIFO order or according to some symmetric scheduling policy, it is often desirable to discriminate among the different job types,

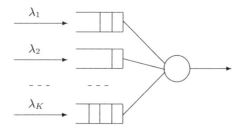

Fig. 3.9. A multiclass M/G/1 system with priorities

giving a better quality of service to some, at the expense of others. The usual mechanism for doing that is to operate some sort of priority policy.

Assume that there are K job types, numbered $1, 2, \ldots, K$. Type i jobs arrive in an independent Poisson process with rate λ_i; their service requirements have some general distribution, with mean b_i and second moment M_{2i} ($i = 1, 2, \ldots, K$). There is a separate unbounded queue for each type, where jobs wait in order of arrival. Service is provided by a single server (figure 3.9).

The different queues are served according to a priority assignment which we shall assume, for convenience, to be in inverse order of type indices. Thus, type 1 has the highest priority, type 2 the second highest, ..., type K the lowest. Whenever a scheduling decision has to be made as to which job to serve next, the job selected is the one at the head of the highest priority (lowest index) non-empty queue. This means, of course, that a type i job may start service only if queues $1, 2, \ldots, i-1$ are empty ($i = 2, 3, \ldots, K$).

In order to complete the definition of the scheduling policy, it remains to specify what happens if a higher priority job arrives and finds a lower priority one in service. One possibility is to take no action other than place the new arrival in its queue and await the scheduling decision that will be made on completing the current service. The priorities are then said to be 'non-preemptive', or 'head-of-the-line'. Alternatively, the new arrival may be allowed to interrupt the current service and occupy the server immediately. The displaced job goes back to the head of its queue. That is a 'preemptive' priority policy.

Preemptive policies are further distinguished by the way they deal with an interrupted job. If the service is eventually continued from the point of interruption, the policy is 'preemptive-resume'. If exactly

3.4 Priority scheduling

the same service is restarted from the beginning, then the policy is 'preemptive-repeat without resampling'. If the job starts a new random service, then the strategy is 'preemptive-repeat with resampling'.

We shall consider the non-preemptive priority policies, and also the preemptive-resume ones, in the steady state. In both cases, the condition for stability is that the total load should be less than 1: $\rho_1+\rho_2+\ldots+\rho_k < 1$. Assume that this is the case.

The following notation will be used:

$\rho_i = \lambda_i b_i$ = offered load of type i;
L_i = average number of jobs of type i in the system;
l_i = average number of jobs of type i waiting for service;
W_i = average response time for a job of type i;
w_i = average total waiting time for a job of type i.

These definitions, and Little's theorem, imply the following relations:

$$L_i = \lambda_i W_i \;\; ; \;\; l_i = \lambda_i w_i \;\; ; \;\; W_i = w_i + b_i \;\; ; \;\; i = 1, 2, \ldots, K . \quad (3.46)$$

Concentrate now on the non-preemptive priority policy. Tag an incoming job of type i and examine the time it has to wait before starting service. That delay can be represented as a sum of three distinct components:

$$w_i = w_0 + A_i + H_i , \quad (3.47)$$

where w_0 is the residual service time of the job that may be found in service; A_i is the delay due the jobs of equal or higher priority (i.e. types $1, 2, \ldots, i$) that are already waiting when the tagged job arrives; and H_i is the additional delay resulting from the higher priority jobs (types $1, 2, \ldots, i-1$) that arrive while the tagged job is waiting.

According to the utilization law (and PASTA), the tagged job finds a job of type j in service with probability ρ_j. If a type j service is in progress, its average residual lifetime is $M_{2j}/(2b_j)$ (section 2.1.1). Hence,

$$w_0 = \sum_{j=1}^{K} \rho_j \frac{M_{2j}}{2b_j} = \frac{1}{2} \sum_{j=1}^{K} \lambda_j M_{2j} . \quad (3.48)$$

When the tagged job arrives, it finds, on the average, l_j jobs waiting in queue j. Since each of them takes an average of b_j to serve, the total average delay due to the waiting jobs is

$$A_i = \sum_{j=1}^{i} l_j b_j = \sum_{j=1}^{i} \rho_j w_j \;\; ; \;\; i = 1, 2, \ldots, K . \quad (3.49)$$

Finally, during the waiting time of the tagged job, an average of $\lambda_j w_i$ jobs of type j arrive into the system. Each of them requires an average service time of b_j. Therefore, the total additional delay due to subsequent higher priority arrivals is given by

$$H_i = \sum_{j=1}^{i-1} \lambda_j w_i b_j = w_i \sum_{j=1}^{i-1} \rho_j \; ; \; i = 2, 3, \ldots, K \; . \tag{3.50}$$

Of course, the component H_i does not exist for $i = 1$.

Putting together (3.47), (3.48), (3.49) and (3.50), we obtain a set of simultaneous equations for w_i:

$$w_i = w_0 + \sum_{j=1}^{i} \rho_j w_j + w_i \sum_{j=1}^{i-1} \rho_j \; , \tag{3.51}$$

where the last term is 0 for $i = 1$. This set is triangular and therefore quite easily solved by successive substitutions. Its solution is

$$w_i = \frac{w_0}{(1 - \sigma_{i-1})(1 - \sigma_i)} \; ; \; i = 1, 2, \ldots, K \; , \tag{3.52}$$

where w_0 is given by (3.48), and σ_i is the total offered load of priority higher than or equal to i:

$$\sigma_i = \sum_{j=1}^{i} \rho_j \; ; \; i = 1, 2, \ldots, K \; ; \; \sigma_0 = 0 \; .$$

Expressions (3.52) are due to A. Cobham and bear his name. The average response times, W_i, are given by

$$W_i = \frac{w_0}{(1 - \sigma_{i-1})(1 - \sigma_i)} + b_i \; ; \; i = 1, 2, \ldots, K \; . \tag{3.53}$$

The average numbers of type i jobs in the system and in the queue, L_i and l_i, are obtained from (3.46).

Certain implications of the above derivation are worth pointing out. First, the average waiting (and response) time for type i is influenced by the lower priority types $(i + 1, i + 2, \ldots, K)$ only through the expected residual service delay, w_0. It is obvious from the way that quantity was obtained, that it is in fact independent of the scheduling policy, provided that the latter does not allow service interruptions. The higher priority job types $(1, 2, \ldots, i - 1)$ influence type i through their total offered load σ_{i-1}, as well as through w_0. That total load is also independent of the scheduling policy.

Thus, as far as jobs of type i are concerned, all jobs can be separated into three essential groups: group 1, which comprises the higher priority types $1, 2, \ldots, i - 1$; group 2, which has type i on its own; and group 3, which consists of the lower priority types $i + 1, i + 2, \ldots, K$. The order of service within groups 1 and 3 is immaterial. Any non-preemptive scheduling policy can be employed there (e.g. FIFO), without affecting type i.

A look at the denominator in (3.52) convinces us that, in order for w_i to be finite, it is sufficient that the total load for types $1, 2, \ldots, i$ is less than 1. Suppose, for instance, that $\sigma_i < 1$, but $\sigma_{i+1} > 1$. Then the first i queues are stable, while queue $i + 1$ (and all lower priority ones) grow without bound as time increases. The probability that a job of type $i + 1$ is in service is equal to $1 - \sigma_i$. Jobs of types $i + 2, \ldots, K$ are never served, in the long run. Cobham's expressions continue to hold, up to and including type i. However, the formula (3.48) for w_0 should be replaced by

$$w_0 = \frac{1}{2} \left[\sum_{j=1}^{i} \lambda_j M_{2j} + (1 - \sigma_i) \frac{M_{2,i+1}}{b_{i+1}} \right].$$

The average waiting times for types $i + 1, i + 2, \ldots, K$ are infinite.

Example

1. The SPT policy. Jobs arrive into a single-server system according to a Poisson process with rate λ. Their service times, X, have K possible values: $X = x_i$ with probability f_i ($i = 1, 2, \ldots, K$; $f_1 + f_2 + \ldots + f_K = 1$), where $x_1 < x_2 < \cdots < x_K$. These required service times are known on arrival and the scheduling policy gives higher non-preemptive priority to shorter jobs (jobs of length x_1 have top priority, ..., those of length x_K have bottom priority). This policy is called 'Shortest-Processing-Time-first', or SPT.

When is the system stable and what is the average waiting time for the jobs that require service x_i ($i = 1, 2, \ldots, K$)?

Here we have a priority model where the type of a job is determined by its length, or required service time. From the decomposition property of the Poisson process (section 2.3.1) it follows that jobs of type i arrive in a Poisson process with rate λf_i. The first and second moments of their service times are, of course, x_i and x_i^2, respectively. The total load corresponding to types $1, 2, \ldots, i$ is

$$\sigma_i = \lambda \sum_{j=1}^{i} f_j x_j \; ; \; i = 1, 2, \ldots, K \, .$$

In order that the system may be stable, the total load from all types should be less than 1:

$$\sigma_K = \lambda E(X) < 1 \, .$$

When the system is in the steady state, the average residual service delay is equal to

$$w_0 = \frac{\lambda}{2} \sum_{i=1}^{K} f_i x_i^2 = \frac{\lambda M_2}{2} \, ,$$

where M_2 is the second moment of the required service time.

The average waiting time for jobs of type i, $w(x_i)$, is given by Cobham's formula:

$$w(x_i) = \frac{\lambda M_2}{2(1 - \sigma_{i-1})(1 - \sigma_i)} \; ; \; i = 1, 2, \ldots, K \, .$$

3.4.1 Preemptive-resume priorities

Consider now the steady-state performance of the preemptive-resume priority scheduling policy. The service of a lower priority job may be interrupted by the subsequent arrival of a higher priority one. The preempted job is returned to the head of its queue and eventually resumes its service from the point of interruption. This may happen several times before the service is completed.

Define the 'initial waiting time' of a tagged job of type i as the period between its arrival and the start of its service. That is followed by what we shall call the 'attendance time', which is the period between the start and the completion of the service. The attendance time may contain further waits caused by preemptions. Denote the averages of the initial waiting time and the attendance time by v_i and s_i, respectively.

The following two observations will help us to determine v_i. First, the jobs of lower priority types, $i + 1$, $i + 2$, ..., K, may be ignored completely. They can be preempted, and therefore have no influence on the performance of type i. Second, the initial waiting time of the tagged job does not depend on whether the scheduling policy among types $1, 2, \ldots, i - 1$ is preemptive or not. In either case, it can start service only when it reaches the head of its queue and all higher priority queues are empty.

We conclude, therefore, that v_i is equal to the average waiting time of a type i job under a *non-preemptive* priority policy, in a system where types $i+1, i+2, \ldots, K$ do not exist. The corresponding Cobham formula is

$$v_i = \frac{w_{0i}}{(1 - \sigma_{i-1})(1 - \sigma_i)}, \qquad (3.54)$$

where w_{0i} is the residual service delay involving types $1, 2, \ldots, i$ only:

$$w_{0i} = \frac{1}{2} \sum_{j=1}^{i} \lambda_j M_{2j} \;\; ; \;\; i = 1, 2, \ldots, K \;.$$

Next, the attendance time of the tagged job consists of its own service time, plus the service times of all higher priority jobs that arrive during the attendance time. An average of $\lambda_j s_i$ jobs of type j arrive during an interval of average length s_i; each of them takes an average of b_j to serve. Hence, we can write

$$s_i = b_i + \sum_{j=1}^{i-1} \lambda_j s_i b_j = b_i + s_i \sigma_{i-1} \;\; ; \;\; i = 1, 2, \ldots, K \;, \qquad (3.55)$$

where $\sigma_0 = 0$. Solving this for s_i and combining it with (3.54) yields the average response time for type i:

$$W_i = \frac{w_{0i}}{(1 - \sigma_{i-1})(1 - \sigma_i)} + \frac{b_i}{(1 - \sigma_{i-1})} \;\; ; \;\; i = 1, 2, \ldots, K \;. \qquad (3.56)$$

The total waiting time is of course $w_i = W_i - b_i$, while the average numbers of jobs, L_i and l_i, are obtained from (3.46).

3.4.2 Averaging and lumping job types

The criterion for evaluating the performance of a scheduling policy in a multiclass system is often the overall, or unconditional, average response time. This is easily derived from the type-dependent, or conditional, performance measures. The total average number of jobs in the system, L, is a straightforward sum

$$L = \sum_{i=1}^{K} L_i \;. \qquad (3.57)$$

3 Queueing systems: average performance

Let $\lambda = \lambda_1 + \lambda_2 + \ldots + \lambda_K$ be the total arrival rate. Little's result, applied to the whole system, yields the overall average response time, W:

$$W = \frac{L}{\lambda} = \frac{1}{\lambda}\sum_{i=1}^{K} L_i = \frac{1}{\lambda}\sum_{i=1}^{K} \lambda_i W_i \; . \tag{3.58}$$

The last equation has a simple probabilistic interpretation. Since an average of λ_i jobs of type i arrive into the system per unit time, λ_i/λ is the fraction of all arrivals that are of type i, or the probability that a new arrival is of type i. The right-hand side of (3.58) is thus the appropriate expression for obtaining an unconditional average from conditional ones.

It may be of interest to compare the system performance under a priority policy, with the performance that would be achieved by lumping together all jobs, putting them in a single queue and serving them in FIFO order, regardless of type. The service time distribution in the resulting M/G/1 queue is a linear combination of the K original service time distributions, with mean and second moment given by

$$b = \frac{1}{\lambda}\sum_{i=1}^{K} \lambda_i b_i \; ,$$

$$M_2 = \frac{1}{\lambda}\sum_{i=1}^{K} \lambda_i M_{2i} \; .$$

The corresponding performance measures are provided by (3.16):

$$w = \frac{\lambda M_2}{2(1-\sigma_K)} \; ; \; W = w + b \; ; \; W_i = w + b_i \; , \tag{3.59}$$

where σ_K is the total offered load of the K job types. The average FIFO performance may be better, or worse, than the average overall performance under a priority policy. The outcome of the comparison depends on the particular priority assignment; the effect of the latter will be examined in the next section.

Exercises

1. The jobs in a single-processor computer system are of two types. They arrive in Poisson processes, with rates 5 and 10 jobs per minute, respectively. The required service times for both types are exponentially distributed, with means 6 seconds and 2 seconds, respectively.

Find the average response times for the two job types, and the overall average response time, under the following scheduling policies:

(a) type 1 has non-preemptive top priority;
(b) type 1 has preemptive-resume top priority;
(c) type 2 has non-preemptive top priority;
(d) type 2 has preemptive-resume top priority;
(e) all jobs join a single queue and are served in FIFO order.

2. Generalize the SPT model from example 1 in section 3.4 by assuming that the service times have some general distribution function, $F(x)$. Required service times are known exactly on arrival and shorter jobs have higher priority. Show that the expected residual service delay, w_0, is again equal to $\lambda M_2/2$, where M_2 is the second moment of the service time. Using the required service time of a job as a type identifier and replacing the sums in Cobham's expressions by integrals, derive the following formula for the average waiting time, $w(x)$, of a job whose required service time is x:

$$w(x) = \frac{w_0}{[1 - \lambda \int_0^{x-} u dF(u)][1 - \lambda \int_0^{x} u dF(u)]},$$

where $x-$ indicates limit from the left. If the distribution function $F(\cdot)$ is continuous at point x, then the two terms in square brackets are equal.

3.5 Optimal scheduling policies

The factors that influence the performance of a queueing system can be grouped into three broad categories. These are (i) the hardware characteristics (e.g. server speed), (ii) the demand characteristics (e.g. job types, arrival rates, service requirements) and (iii) the scheduling policy employed. The first two categories are usually difficult to control, but the scheduling policy may be easily changed. One is then faced with the problem of selecting the best policy for the particular system parameters.

We shall begin by addressing that question first in the context of the family of non-preemptive priority strategies. The assumptions are as in the previous section. The demand consists of K job types, arriving in independent Poisson processes and having general service requirements. The parameters for type i are: arrival rate λ_i, average service time b_i and second moment of service time M_{2i}.

There is a choice of $K!$ policies: each permutation of job type indices,

$(i_1, i_2, \ldots, i_K)$, specifies a possible priority assignment where type i_1 has top priority, type i_2 has second top priority, etc. In the last section, the priority assignment was defined by the permutation $(1, 2, \ldots, K)$.

The object is to minimize a cost function of the form

$$C = \sum_{i=1}^{K} c_i w_i , \qquad (3.60)$$

where w_i is the average waiting time for jobs of type i, and c_i is a non-negative constant. The latter reflects the relative importance attached to type i. We could have chosen as a cost function a linear combination of average response times, rather than waiting times. However, since $W_i = w_i + b_i$, the two objectives are equivalent.

Intuitively, if c_i is large, then in order to keep the cost low, w_i should be small, i.e. type i should be given high priority. However, a priority ordering based solely on the magnitudes of the coefficients c_i is not necessarily optimal. The loads $\rho_i = \lambda_i b_i$ also play a role in determining the best strategy, since they influence the waiting times.

In principle, one could use Cobham's expressions (3.52) to evaluate the cost function for each of the $K!$ priority assignments, and then compare the results. However, such an approach is impractical even for moderately large values of K. Fortunately, the optimal scheduling strategy can be found without performing a search.

First, we shall establish a relation between the K average response times which, besides being the key to our argument, is of interest in its own right.

3.5.1 A conservation law for waiting times

Certain performance characteristics of multiclass single-server systems are independent of the scheduling policy. We have already encountered two of them: the server utilization and the busy period. Now we shall introduce a third such quantity. The 'virtual load', $V(t)$, is defined as the total unfinished amount of work present in the system at time t. In other words, $V(t)$ is the sum of the remaining service requirements of all jobs in the system at time t. If the scheduling policy does not create or lose work, i.e. if the server serves a job whenever any are present, and jobs do not depart before their service is completed, then the virtual

3.5 Optimal scheduling policies

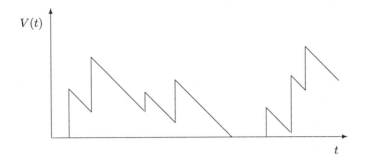

Fig. 3.10. Virtual load in a single server system

load behaves as follows:

1. At every arrival instant, $V(t)$ jumps up by an amount equal to the service requirement of the incoming job.
2. At other times, if $V(t) > 0$ it decreases with gradient -1; when it reaches 0, it remains 0 until the next arrival instant.

A typical realization of a virtual load function is illustrated in figure 3.10.

The above definition implies that $V(t)$ is completely determined by the sequence of arrival instants and the corresponding service requirements. It does not depend on the scheduling policy.

Denote by v the average virtual load in the steady state. In the case of non-preemptive scheduling policies, v is equal to the residual service time of the job that may be in service (w_0 on the average), plus the service times of all waiting jobs (an average of l_i jobs of type i are waiting, and each of them has an average service requirement b_i). That is,

$$v = w_0 + \sum_{i=1}^{K} l_i b_i = w_0 + \sum_{i=1}^{K} \rho_i w_i , \qquad (3.61)$$

where w_0 is given by (3.48), and $l_i = \lambda_i w_i$, by Little's theorem. In this equation, both v and w_0 are independent of the scheduling policy. Therefore, the linear combination of average waiting times that appears in the right-hand side is also independent of the scheduling policy. This result is known as Kleinrock's conservation law.

Theorem 3.2 *(L. Kleinrock) For any single-server queueing system in steady state under a non-preemptive scheduling policy, there is a constant*

v (independent of the policy), such that

$$\sum_{i=1}^{K} \rho_i w_i = v - w_0 \ . \tag{3.62}$$

Note that this result does not require the arrival processes to be Poisson. The only important assumption is that the system is in equilibrium.

To find the value of the constant $v - w_0$, it suffices to determine the average virtual load under any non-preemptive policy that can be easily analysed. For instance, when the arrival processes are Poisson, (3.59) provides an expression for v under the FIFO policy (then the average virtual load is equal to the average waiting time):

$$v - w_0 = w - w_0 = \frac{\lambda M_2 \sigma_K}{2(1 - \sigma_K)} \ , \tag{3.63}$$

where σ_K is the total offered load.

So, any attempt to reduce the waiting times of one or more job types (e.g. by giving them higher priority) results in increased waiting times for the other job types, subject to (3.62).

Conservation law for preemptive policies. When the service times are distributed exponentially, Kleinrock's theorem holds for all scheduling policies, including those that allow preemptions. This is because of the memoryless property: the average remaining service time of any type i job in the queue is b_i, regardless of how much service it has received already. Thus, (3.61) is still valid and so is (3.62).

3.5.2 The c/ρ rule

Let us now examine the way in which the value of the cost function (3.60) changes with the (non-preemptive) priority assignment. Two policies are said to be 'neighbouring' if they differ only in the priority order of two adjacent job types. For instance, in a system with five job types, the policies $(2, 4, 5, 3, 1)$ and $(2, 5, 4, 3, 1)$ are neighbouring. Let A and B be two neighbouring scheduling policies, and i and j be the adjacent job types that swap priorities: $A = (\ldots, i, j, \ldots)$; $B = (\ldots, j, i, \ldots)$. Denote the average waiting time of a type k job under A and B by w_k^A and w_k^B respectively ($k = 1, 2, \ldots, K$). The corresponding values of the cost function are C^A and C^B, respectively.

From the properties of the priority scheduling policies it follows that all job types except i and j have the same average waiting times under

3.5 Optimal scheduling policies

A and B. Hence, the difference $C^A - C^B$ is equal to

$$\begin{aligned} C^A - C^B &= (c_i w_i^A + c_j w_j^A) - (c_i w_i^B + c_j w_j^B) \\ &= c_i(w_i^A - w_i^B) + c_j(w_j^A - w_j^B) . \end{aligned} \quad (3.64)$$

All the other terms cancel out. Also, Kleinrock's conservation law implies that

$$\rho_i w_i^A + \rho_j w_j^A = \rho_i w_i^B + \rho_j w_j^B . \quad (3.65)$$

Again, all the other terms cancel out. This last equality can be rewritten as

$$w_i^A - w_i^B = -\frac{\rho_j}{\rho_i}(w_j^A - w_j^B) . \quad (3.66)$$

After substitution of (3.66), expression (3.64) becomes

$$C^A - C^B = (w_j^A - w_j^B)\left(c_j - c_i \frac{\rho_j}{\rho_i}\right) . \quad (3.67)$$

Note that the first factor in the right-hand side of (3.67) is always positive. This is because type j has higher priority, and hence lower average waiting time, under policy B than under A. Therefore, the sign of the whole expression is determined by that of the second factor. Hence, policy B is better than policy A if, and only if,

$$\frac{c_j}{\rho_j} > \frac{c_i}{\rho_i} . \quad (3.68)$$

It is now easy to identify a priority scheduling policy that minimizes the cost function. Such a policy satisfies the so-called c/ρ rule: if the inequality (3.68) holds for *any* job types i and j, then type j has higher priority than type i. Indeed, if a policy satisfying the c/ρ rule is not optimal, then one of its neighbouring policies, or their neighbouring policies, etc., would be better than it. That, however, is impossible, in view of the above argument.

Suppose that, instead of the cost function (3.60), we wish to minimize a linear combination of the average numbers of type i jobs waiting for service:

$$C = \sum_{i=1}^{K} c_i l_i . \quad (3.69)$$

Little's theorem implies that this cost function is equivalent to (3.60) with coefficients $c_i \lambda_i$. Hence, a policy that minimizes (3.69) is one that satisfies the 'c/b rule': if the inequality $c_j/b_j > c_i/b_i$ holds for any job types i and j, then type j has higher priority than type i.

It turns out that the c/ρ and c/b rules are optimal within the set of all non-preemptive policies, not just the $K!$ non-preemptive priority ones. In other words, any non-preemptive policy which is not based on a priority assignment (e.g. FIFO, LIFO), is bound to be sub-optimal. This will be discussed at the end of the chapter.

Example

1. Minimize the overall average waiting time. Suppose that the coefficients of the cost function are (or are proportional to) the arrival rates of the corresponding job types:

$$C = \sum_{i=1}^{K} \lambda_i w_i \ .$$

Remembering that an incoming job is of type i with probability λ_i/λ, where λ is the total arrival rate, we see that minimizing this cost function is equivalent to minimizing the overall average waiting time, w.

Since $\lambda_i/\rho_i = 1/b_i$, the c/ρ rule suggests that an optimal priority assignment is one which gives higher priority to the job type with lower average service time. That scheduling policy is called 'Shortest-Expected-Processing-Time-first', or SEPT. According to the c/b rule, SEPT also minimizes the total average number of waiting jobs (and jobs in the system).

3.5.3 Optimization with preemptions

We have seen that, in the non-preemptive case, optimal scheduling decisions can be taken without worrying about the distributions of the required service times. Only the means are important. However, if preemptive policies are also allowed, the situation becomes considerably more complicated.

Suppose that, at time 0, there are K jobs at the server, and that the required service time of job i has distribution function $F_i(x)$ ($i = 1, 2, \ldots, K$). There are no further arrivals and services can be interrupted at arbitrary points. The object is to find a scheduling policy that

3.5 Optimal scheduling policies

minimizes a cost function of the form

$$C = \sum_{i=1}^{K} c_i w_i \,, \tag{3.70}$$

where w_i is the total average waiting time of job i.

The solution to this problem was found fairly recently. It is not an intuitively obvious one and is quite difficult to establish rigorously. We shall present the relevant results without proof.

Suppose that job i has already received service x and that the server is assigned to it for a further period y. The conditional probability that the job will complete within that period is

$$G_i(x, y) = \frac{F_i(x+y) - F_i(x)}{1 - F_i(x)} \,. \tag{3.71}$$

The average amount of service that the job will actually use during the period y is equal to

$$S_i(x, y) = \frac{1}{1 - F_i(x)} \int_0^y [1 - F_i(x+u)] du \,. \tag{3.72}$$

If the completion of a job is regarded as a desirable occurrence which is rewarded, then the ratio $G_i(x,y)/S_i(x,y)$ can be thought of as the expected reward from job i per unit of processor time used during the period y. In general, that ratio varies with y. Let $\xi_i(x)$ be the largest value that it can reach, over all possible y:

$$\xi_i(x) = \max_{y > 0} \frac{G_i(x, y)}{S_i(x, y)} \,. \tag{3.73}$$

The quantity $\xi_i(x)$ is called the 'Gittins index' of job i when the latter has attained service x. The smallest value of y for which the maximum in (3.73) is reached is called the 'service quantum'. The optimal scheduling policy can now be described as follows.

At each scheduling decision instant (the first being at time 0), select the job for which the product $c_i \xi_i(x)$ is largest. Assign the server to that job for the duration of the corresponding service quantum. The end of the quantum, or the completion of the job if the latter occurs earlier, defines the next decision instant. Clearly, at most one index has to be re-computed then: the one of the job that had the server (if it has not completed). The attained service times of all other jobs remain unchanged.

This is known as the 'index' scheduling policy. It can be shown that if, instead of being all present at the beginning, jobs of different types arrive in independent Poission processes, the optimal scheduling policy is still, essentially, an index policy. Consider again the model with K job types characterized by their arrival rates, λ_i, and service time distribution functions, $F_i(x)$ ($i = 1, 2, \ldots, K$). The cost function to be minimized is (3.60). Arbitrary preemptions are allowed.

The optimal scheduling policy is then an index policy modified as follows:

1. Scheduling decisions are made at job arrival instants, job completion instants and service quanta terminations.
2. The job selected for service at each such instant is the one for which the quantity $(c_i \xi_i(x)/\lambda_i)$ is largest, where i is the job type, x is its attained service time and $\xi_i(x)$ is the Gittins index computed according to (3.73).

Note the analogy between the c/ρ rule and the index policy. The role of the average service time, b_i, is now played by $1/\xi_i(x)$. Of course, the latter may change during the residence of a job in the system, as it attains portions of its required service.

The index policy has a simple form in special cases. These depend on the following quantities:

$$r_i(x) = \frac{f_i(x)}{1 - F_i(x)} \quad ; \quad i = 1, 2, \ldots, K , \qquad (3.74)$$

where $f_i(x)$ is the pdf of the type i required service time. The right-hand side of (3.74) represents the completion rate of a type i job which has attained service x.

If $r_i(x)$ is a monotone increasing function of x, then any allocation of the server to a type i job will have the effect of increasing its index. Hence, the maximum in (3.73) is reached for $y = \infty$. An infinite service quantum implies that the service of a type i job can be interrupted only by a new arrival with a higher index, and not by any job already in the system. If this is true for all job types, then the index policy operates the c/ρ rule with preemptions, where b_i is replaced by the expected *remaining* service time of a type i job with attained service x.

If $r_i(x)$ is a decreasing function of x, then giving more service to a type i job decreases its index. The maximum in (3.73) is reached for $y = 0$. In that case, $\xi_i(x) = r_i(x)$. The effect of an infinitesimal service quantum is to serve the job with the highest index until the latter's value drops to the level of the next highest one. As soon as there are at least two jobs with equal (highest) indices, the policy becomes Processor-Sharing between them.

The completion rate tends to be an increasing (decreasing) function of x when the corresponding coefficient of variation of the required service time is small (large). This, together with the above observations, leads us to a rather general scheduling principle: if the required service times are reasonably predictable, then having decided to serve a job, it should not be interrupted too much. If, on the other hand, the service times are very variable, frequent interruptions are better. Another manifestation of the same principle was observed in the comparison between the FIFO and Processor-Sharing policies in section 3.3.3.

If all service times are distributed exponentially, then the completion rates are constant and the Gittins indices do not depend on the attained service times. The optimal index policy then coincides with the preemptive priority policy determined by the c/ρ rule.

Example

2. The SRPT policy. The arrival rate in an M/G/1 system is λ and the required service time of every job is known on arrival. Consider for simplicity the discrete case, where the possible service times are $x_1, x_2, \ldots$, with probabilities $f_1, f_2, \ldots$ respectively. A job is of type i if its service time is x_i. Thus, the arrival rate for type i is λf_i and the service time distribution function is

$$F_i(x) = \begin{cases} 0 & \text{for } x < x_i \\ 1 & \text{for } x \geq x_i \end{cases}.$$

The definitions (3.71) and (3.72) imply that the Gittins index of a type i job which has received service x ($x < x_i$) is equal to the reciprocal of the remaining service time:

$$\xi_i(x) = \frac{1}{x_i - x}.$$

The corresponding service quantum is $x_i - x$. To minimize the cost function (3.70), one should give higher preemptive priority to jobs with higher value of $c_i/[f_i(x_i - x)]$. If the objective is to minimize the overall average waiting time (i.e. c_i is proportional to f_i), then the optimal policy serves shorter jobs first, and if a new arrival has a shorter service than the residual of the service in progress, the latter is preempted. This is the Shortest-Remaining-Processing-Time-first policy (SRPT).

3.5.4 Characterization of achievable performance

A problem which is closely related to optimization concerns the characterization of achievable performance. Suppose, for instance, that in our single-server system with K job types, performance is measured by the vector of average waiting times for the different types, $\mathbf{w} = (w_1, w_2, \ldots, w_K)$. What is the set of such vectors that can be achieved by varying the scheduling policy? Or, to put it another way, if a given vector of waiting times is specified as a performance objective, is there a scheduling policy that can achieve it?

The general characterization problem is still open. There are, however, solutions in the cases where Kleinrock's conservation law applies: either the scheduling policies are non-preemptive and the service time distributions are general, or the policies are general and the distributions are exponential. Let us examine, as an illustration, the case of two job types with Poisson arrivals, and consider the performance vectors, $\mathbf{w} = (w_1, w_2)$, achievable by non-preemptive scheduling policies. Those vectors can be represented as points on the two-dimensional plane.

The conservation law tells us that all achievable points lie on the straight line defined by

$$\rho_1 w_1 + \rho_2 w_2 = \frac{(\lambda_1 M_{21} + \lambda_2 M_{22})(\rho_1 + \rho_2)}{2(1 - \rho_1 - \rho_2)} . \tag{3.75}$$

There are two special points on that line, $\mathbf{w}_{12}$ and $\mathbf{w}_{21}$, corresponding to the two non-preemptive priority assignments (1,2) and (2,1) (see figure 3.11).

Those two points (whose coordinates are obtained by means of Cobham's expressions, section 3.4) are the extremes of the set of achievable performance vectors. The priority policy $(1, 2)$ yields the lowest possible (without preemptions) average response time for type 1 and the highest possible one for type 2; the situation is reversed with the policy $(2, 1)$. Thus, no point to the left of $(1, 2)$, or to the right of $(2, 1)$, can be achieved.

On the other hand, it can be seen that all points between the two extremes are achievable. Indeed, one can construct a 'mixed' policy which, at the beginning of each busy period, chooses to operate the priority assignment $(1,2)$ with probability α, and the assignment $(2,1)$ with probability $1 - \alpha$. The performance vector of that mixed policy, $\mathbf{w}_\alpha$, is

$$\mathbf{w}_\alpha = \alpha \mathbf{w}_{12} + (1 - \alpha) \mathbf{w}_{21} .$$

3.5 Optimal scheduling policies

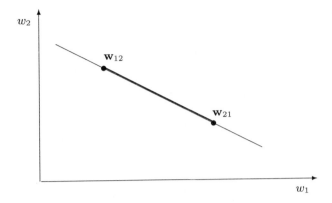

Fig. 3.11. Achievable performance vectors in a 2-class M/G/1 system

Clearly, when α varies between 0 and 1, the point $\mathbf{w}_\alpha$ sweeps the entire line segment between $\mathbf{w}_{21}$ and $\mathbf{w}_{12}$.

In the case of K job types, all achievable performance vectors lie in a $(K-1)$-dimensional hyperplane defined by the conservation law. There are $K!$ extreme points, corresponding to the $K!$ non-preemptive priority policies. The set of achievable performance vectors is then the polyhedron which has those points as its vertices (figure 3.12 illustrates the case $K=3$).

This characterization implies that the maximum, or minimum, of any cost function which is linear in the average waiting times is reached at one of the vertices. Consequently, one of the $K!$ non-preemptive scheduling policies is guaranteed to be optimal. The appropriate priority assignment is determined by the c/ρ rule.

When the scheduling policies are general and the service time distributions are exponential, the set of achievable performance vectors coincides with the $(K-1)$-dimensional polyhedron whose vertices are the performance vectors of the $K!$ preemptive-resume priority policies. Any linear cost function is optimized by one of those policies.

Exercises

1. Three types of jobs arrive into a system in independent Poisson streams and are served by a single server. Their parameters are: (type 1) arrival rate 0.2, constant service times of length 1; (type 2) arrival

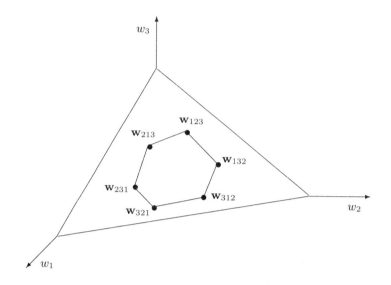

Fig. 3.12. Achievable performance vectors in a 3-class M/G/1 system

rate 0.1, service times uniformly distributed between 0 and 10; (type 3) arrival rate 2, service times distributed exponentially with mean 0.05. Let W_i and W be the type i average response time ($i = 1, 2, 3$), and the overall average response time, respectively, corresponding to a non-preemptive scheduling policy.

(a) Find a scheduling policy that minimizes W.
(b) Find a scheduling policy that minimizes the cost function $3W_1 + 2W_2 + W_3$.
(c) Find a scheduling policy that minimizes W_2.
(d) Calculate the value of W_2 achieved by the policy found in (c).

2. Jobs arrive into a single-server system in a Poisson process with rate λ; the distribution function of their service requirements is $F(x)$. The exact required service time of every job is known on arrival. The object is to minimize the overall average waiting (or response) time, using a non-preemptive scheduling policy.

Treating the required service time as a type identifier and applying the result from example 1 in section 3.5.2, show that the policy which gives non-preemptive priority to shorter jobs over longer ones is optimal.

For the performance of the SPT policy, see example 1 and exercise 2 in section 3.4.

3. In a single-server system with K job types and Poisson arrivals, the required service times for type i are uniformly distributed on the interval $(0, d_i)$ ($i = 1, 2, \ldots, K$). Show that the index of a type i job with attained service time x is equal to

$$\xi_i(x) = \frac{2}{d_i - x} \ ; \ 0 < x < d_i \ , \ i = 1, 2, \ldots, K \ ,$$

and that the corresponding service quantum is $y = d_i - x$. Hence establish that a policy that minimizes the overall average response time is the preemptive 'Shortest-Expected-Remaining-Processing-Time first' policy (SERPT).

4. In a single-server system with K job types and Poisson arrivals, the required service times for type i have a 'hyperexponential' distribution:

$$F_i(x) = 1 - \alpha_i e^{-\mu_i x} - (1 - \alpha_i)e^{-\nu_i x} \ ,$$

where $0 < \alpha_i < 1$, and μ_i and ν_i are positive parameters.

Show that the completion rate $r_i(x)$ is a decreasing function of x, for all job types. Hence determine a scheduling policy (which may use preemptions) that minimizes a linear cost function.

3.6 Literature

The first proof of Little's theorem appeared in [13]. The arguments in this chapter are due to Foster [5].

Listed below are several books containing queueing theory results and applications [2,8,11,12,14]. A very extensive bibliography on the subject can be found in Dshalalow [4], which also presents a collection of more recent developments. Two useful references on priority scheduling policies are Conway et al. [3], and Jaiswal [9]. The first of these presents a unified treatment of all priority policies (including preemptive-repeat with and without resampling), based on busy period analysis. The original studies of the M/M/1 and M/G/1 Processor-Sharing systems are by Coffman et al. [1] and Yashkov [17].

The class of symmetric scheduling policies was introduced by Kelly [10].

The subject of optimal scheduling, both deterministic and stochastic, has received much attention. The optimality of the index policy was discovered independently by Gittins and Jones [7], and by Sevcik [15]. A comprehensive coverage of that area is provided by Whittle [16]. Discussion and proofs of conservation laws, and of several achievability results, can be found in Gelenbe and Mitrani [6].

References

1. E.G. Coffman, R.R. Muntz and H. Trotter, "Waiting Time Distributions for Processor-Sharing Systems", *JACM*, **17**, 123–130, 1970.
2. J.W. Cohen, *The Single Server Queue*, Wiley-Interscience, 1969.
3. R.W. Conway, W.L. Maxwell and L.W. Miller, *Theory of Scheduling*, Addison-Wesley, 1967.
4. J.H. Dshalalow (editor) *Advances in Queueing*, CRC Press, 1995.
5. F.G. Foster, "Stochastic Processes", Procs., IFORS Conference, Dublin, 1972.
6. E. Gelenbe and I. Mitrani, *Analysis and Synthesis of Computer Systems*, Academic Press, 1980.
7. J.C. Gittins and D.M. Jones, "A Dynamic Allocation Index for the Sequential Design of Experiments", in *Progress in Statistics* (edited by J. Gani), North-Holland, 1974.
8. P.G. Harrison and N.M. Patel, *Performance Modelling of Communication Networks and Computer Architectures*, Addison-Wesley, 1993.
9. N.K. Jaiswal, *Priority Queues*, Academic Press, 1968.
10. F.P. Kelly, *Reversibility and Stochastic Networks*, John Wiley, 1979.
11. P.J.B. King, *Computer and Communication Systems Performance Modelling*, Prentice-Hall, 1990.
12. L. Kleinrock, *Queueing Systems*, Volumes 1 and 2, John Wiley, 1975, 1976.
13. J.D.C. Little, "A Proof for the Queueing Formula $L = \lambda W$", *Operations Research*, **9**, 1961.
14. T.L. Saaty, *Elements of Queueing Theory and its Applications*, McGraw-Hill, 1961.
15. K.C. Sevcik, "A Proof of the Optimality of Smallest Rank Scheduling", *JACM*, **21**, 1974.

16. P. Whittle, *Optimization Over Time*, Volumes 1 and 2, Wiley, 1982, 1983.
17. S.F. Yashkov, "A Derivation of Response Time Distribution for an M/G/1 Processor-Sharing Queue", *Probability, Control and Information Theory*, **12**, 133–148, 1983.

4

Queueing networks

Some of the most important applications of probabilistic modelling techniques are in the area of distributed systems. The term 'distributed' means, in this context, that various tasks that are somehow related can be carried out by different servers which may or may not be in different geographical locations. Such a broad definition covers a great variety of applications, in the areas of manufacturing, transport, computing and communications. To study the behaviour of a distributed system, one normally needs a model involving a number of service centres, with jobs arriving and circulating among them according to some random or deterministic routeing pattern. This leads in a natural way to the concept of a network of queues.

A queueing network can be thought of as a connected directed graph whose nodes represent service centres. The arcs between those nodes indicate one-step moves that jobs may make from service centre to service centre (the existence of an arc from node i to node j does not necessarily imply one from j to i). Each node has its own queue, served according to some scheduling strategy. Jobs may be of different types and may follow different routes through the network. An arc without origin leading into a node (or one without destination leading out of a node) indicates that jobs arrive into that node from outside (or depart from it and leave the network). Figure 4.1 shows a five-node network, with external arrivals into nodes 1 and 2, and external departures from nodes 1 and 5. At this level of abstraction, only the connectivity of the nodes is specified; nothing is said about their internal structure, nor about the demands that jobs place on them.

In order to complete the definition of a queueing network model, one has to make assumptions about the nature of the external arrival streams, the routeing of jobs among nodes and, for each node, the num-

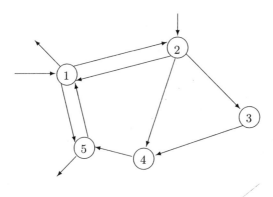

Fig. 4.1. A queueing network with five nodes

ber of servers available, the required service times and the scheduling strategy. If these assumptions are sufficiently 'nice', the resulting model can be solved analytically or numerically, and the values of various performance measures can be obtained. In many other cases, where exact results are not available, good approximations can be derived by making appropriate simplifications.

There is a generally accepted classification of queueing networks, depending on (a) the topology of the underlying graph and (b) the nature of the job population. A network is said to be 'open' if there is at least one arc along which jobs enter it and at least one arc along which jobs leave it, and if from every node it is possible to follow a path leading eventually out of the network. In other words, an open network is one in which no job can be trapped and prevented from leaving. Figure 4.1 provides an illustration of an open network. By contrast, a network without external arrivals and without departures, but with a fixed number of jobs circulating forever among the nodes, is called 'closed'. The network in figure 4.1 would become a closed network if the external incoming arcs at nodes 1 and 2, and the outgoing ones from nodes 1 and 5, were removed.

There are also networks which are neither open nor closed. For example, the network in figure 4.2 is not closed because there are external arrivals into node 1 and external departures from node 2. On the other hand, it is not open because any job which visits node 3 can no longer leave. That network, and any other of a similar type, is not stable: the number of trapped jobs keeps increasing without bound.

If all jobs in the network have the same behaviour (statistically), concerning both the services they require and the paths they take among

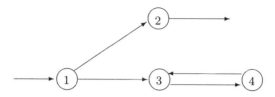

Fig. 4.2. A network which is neither open nor closed

the nodes, then we say that the network has a single job type. In networks with multiple job types, different jobs may have different service requirement characteristics and different routeing patterns. Such networks can also be 'mixed', i.e. open with respect to some job types and closed with respect to others. For example, in figure 4.1 there could be two job types: one using all the arcs, and another (of which there is a fixed number of jobs) endlessly circulating between nodes 1 and 2. The network would then be mixed.

4.1 Open networks

Consider a queueing network consisting of N nodes, numbered $1, 2, \ldots, N$. The simplest set of assumptions that lead to a tractable model are the following:

1. Every node contains a single server and an unbounded queue; the scheduling policy is FIFO.
2. The required service times at node i are distributed exponentially with mean b_i.
3. Jobs arrive *externally* into node i in an independent Poisson process, with rate γ_i.
4. A job completing service at node i goes to node j with probability q_{ij}, regardless of its past history.

The probabilities q_{ij} are called the 'routeing probabilities' of the network; the $N \times N$ matrix $Q = [q_{ij}]$; $i, j = 1, 2, \ldots, N$ is its 'routeing matrix'. Denote by q_i the sum of row i in Q:

$$q_i = \sum_{j=1}^{N} q_{ij} \, . \tag{4.1}$$

4.1 Open networks

All these row sums are less than or equal to 1. Moreover, if $q_i < 1$, then a job completing service at node i leaves the network with probability $1 - q_i$.

In this, and all subsequent queueing network models, it is assumed that the transfers of jobs from node to node are instantaneous. In cases where that assumption appears unreasonable, it is always possible to introduce transfer delays by means of artificial intermediate nodes.

Assumption 3, which postulates independent Poisson external arrivals into different nodes, can be replaced by the following:

3(a). The process of external arrivals into the network is Poisson, with rate γ. Each externally arriving job joins node i with probability α_i.

The decomposition property of the Poisson process implies that 3(a) is equivalent to 3, with $\gamma_i = \alpha_i \gamma$.

Thus the network parameters are the external arrival rates, γ_i, the average service times, b_i, and the routeing matrix, Q. For this network to be open, it is necessary that at least one of the external arrival rates is non-zero and at least one of the row sums q_i is strictly less than 1. Moreover, the matrix Q must be such that from every node there is a sequence of moves leading out of the network with a non-zero probability.

4.1.1 Traffic equations

Suppose that the network is in steady state. Let λ_i be the total average number of jobs arriving into, and departing from, node i per unit time. Some of the incoming jobs may be external (with rate γ_i); others may come from other nodes (including, perhaps, node i itself). On the average, λ_j jobs leave node j per unit time; of these, a fraction q_{ji} go to node i. Therefore, the rate of traffic from node j to node i is $\lambda_j q_{ji}$ ($j = 1, 2, \ldots, N$). Similarly, the rate of traffic from node i to node j is $\lambda_i q_{ij}$. The arrivals and departures at node i are illustrated in figure 4.3.

Expressing the total arrival rate at each node as a sum of external and internal traffic rates, we get a set of linear equations for the unknown quantities λ_i:

$$\lambda_i = \gamma_i + \sum_{j=1}^{N} \lambda_j q_{ji} \ ; \ i = 1, 2, \ldots, N \ . \qquad (4.2)$$

These are known as the 'traffic equations' of the network. Introducing the row vectors $\lambda = (\lambda_1, \lambda_2, \ldots, \lambda_N)$ and $\gamma = (\gamma_1, \gamma_2, \ldots, \gamma_N)$, they can

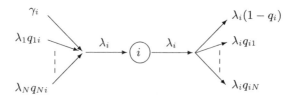

Fig. 4.3. Incoming and outgoing traffic at node i

be written in the form
$$\lambda(I - Q) = \gamma, \qquad (4.3)$$
where I is the identity matrix of order N.

From the definition of an open network it follows that the traffic equations have a unique solution. Indeed, suppose that that is not true, and the matrix $I - Q$ is singular. Then the equations $\lambda(I - Q) = 0$ would have a non-zero solution. In other words, there may be one or more non-zero total arrival rates in the steady state, even if all the external arrival rates are 0. That, however, is impossible: an open network without external input is certain to become empty in the long run. Hence, $I - Q$ is non-singular.

It should be clear from the derivation of the traffic equations that their validity does not depend on the external arrival processes being Poisson. The only important assumption is that of stationarity. Also note that the service times play no role whatsoever in (4.2). However, the condition for stability of the network depends on the offered loads at all nodes, which in turn depend on the parameters b_i. The following must hold:
$$\rho_i = \lambda_i b_i < 1 \; ; \; i = 1, 2, \ldots, N . \qquad (4.4)$$

Thus, we begin the analysis of an open queueing network by saying "If the network is stable, the traffic equations are valid and can be solved." Having solved them, we check the inequalities (4.4); if at least one of them is violated, the network cannot be in steady state and therefore the traffic equations are invalid.

Examples

1. The simplest non-trivial network consists of a single node with feedback, as illustrated in figure 4.4. The external arrival process has rate

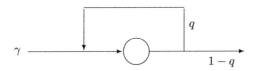

Fig. 4.4. A single-node network with feedback

γ. After completing service, jobs return to the queue with probability q, and leave the system with probability $1-q$.

NB: the branching arcs in the figure are marked with the corresponding routeing probabilities, rather than the traffic rates. This will always be our practice in future.

The routeing matrix of this network consists of a single element, q. There is one traffic equation, which determines the total arrival rate, λ:

$$\lambda = \gamma + \lambda q.$$

This yields $\lambda = \gamma/(1-q)$. If the average service time is b, then the condition for stability of this system is $\gamma b < (1-q)$. The traffic rate along the feedback arc is $\lambda q = \gamma q/(1-q)$. The rate of departures from the network is $\lambda(1-q)$, which is of course equal to the external arrival rate, γ.

2. Consider the four-node network illustrated in figure 4.5. Jobs arrive externally into nodes 1 and 2 with equal rates, $\gamma/2$. The routeing is as follows: from node 1 jobs go to node 3 or 4 with probabilities 0.3 and 0.7 respectively; from node 2 to node 3 or 4 with probabilities 0.7 and 0.3 respectively. From node 3 jobs go back to node 1 with probability 0.8 and leave the network with probability 0.2; similarly, from node 4 they go back to node 2 with probability 0.8 and leave the network with probability 0.2.

The traffic equations for this network are:

$$\lambda_1 = \frac{\gamma}{2} + 0.8\lambda_3 \quad ; \quad \lambda_2 = \frac{\gamma}{2} + 0.8\lambda_4,$$
$$\lambda_3 = 0.3\lambda_1 + 0.7\lambda_2 \quad ; \quad \lambda_4 = 0.7\lambda_1 + 0.3\lambda_2.$$

Adding these equations two by two, we find that

$$\lambda_1 + \lambda_2 = \lambda_3 + \lambda_4 = 5\gamma.$$

Then a simple elimination shows that all arrival rates are in fact equal.

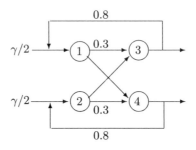

Fig. 4.5. A symmetric four-node network

The solution is:

$$\lambda_1 = \lambda_2 = \lambda_3 = \lambda_4 = 2.5\gamma \ .$$

The average service times can be different, but must all be less than $1/(2.5\gamma)$.

4.1.2 Performance measures

The solution of an open queueing network with Poisson external arrivals, probabilistic routing and exponentially distributed service times turns out to be surprisingly simple. Denote by $p(n_1, n_2, \ldots, n_N)$ the steady-state probability that there are n_1 jobs at node 1, n_2 jobs at node 2, ..., n_N jobs at node N. Those probabilities are given by a famous result known as 'Jackson's theorem'. In the case where every node contains a single server, the latter can be stated as follows:

Theorem 4.1 *(J.R. Jackson) If the arrival rates obtained from the traffic equations are such that $\rho_i < 1$ for all $i = 1, 2, \ldots, N$, then steady state exists and*

$$p(n_1, n_2, \ldots, n_N) = \prod_{i=1}^{N} (1 - \rho_i) \rho_i^{n_i} \ . \qquad (4.5)$$

We shall omit the proof and concentrate on the implications. The fact that the joint distribution of n_1, n_2, ..., n_N can be expressed as a product of N modified geometric distributions is quite remarkable. It implies the following two corollaries, both of which are counter-intuitive.

(a) The marginal distribution of the number of jobs at node i is the same as for an isolated M/M/1 queue with parameters λ_i and b_i (see section 3.2.2).

(b) The numbers of jobs present at any moment in different nodes are independent of each other.

The independence of node states is somewhat unexpected. Intuition might suggest that, since nodes send and receive jobs from other nodes, a long queue at a given node is indicative of long queues at neighbouring nodes. However, that is not so.

The fact that each node behaves like an M/M/1 queue is even more surprising, because the total input is not Poisson, in general. For example, in the simple one-node network in figure 4.4, it can be verified that the total arrival process (composed of the external arrivals and the feedback jobs) is not Poisson. Yet the distribution of the number of jobs in the node is the same as if that process were Poisson.

Generalizations. Jackson's theorem applies to networks where different nodes contain different numbers of parallel servers; apart from this, assumptions 1–4 remain unchanged. If there are k_i servers at node i ($i = 1, 2, \ldots, N$), then the right-hand side of (4.5) is replaced by a product of M/M/k_i distributions (to be derived in chapter 5). In particular, node i may contain infinitely many parallel servers and hence no queue; every job in it is delayed for an average interval b_i, independently of any other jobs. According to the terminology introduced in section 3.1.1, such nodes will be called 'independent delay nodes'. They can be used, for example, to model non-zero transfer times between other nodes.

Jackson's theorem allows us to derive many network performance measures by a straightforward application of existing results. Thus, if node i contains a single server, the average number of jobs in it is given by the M/M/1 formula in section 3.2.2:

$$L_i = \frac{\rho_i}{1 - \rho_i} \ ; \ i = 1, 2, \ldots, N . \qquad (4.6)$$

The average time jobs spend at node i on each visit, W_i, is also given by the corresponding M/M/1 result:

$$W_i = \frac{b_i}{1 - \rho_i} \ ; \ i = 1, 2, \ldots, N . \qquad (4.7)$$

On the other hand, if node i is an independent delay node, then the average time spent in it per visit is $W_i = b_i$, and the average number of jobs in it is $L_i = \rho_i$ (section 3.1.1).

The total average number of jobs in the network, L, is found by summing the averages over all nodes:

$$L = \sum_{i=1}^{N} L_i . \qquad (4.8)$$

Denote by W the total average response time, i.e. the average interval between the arrival of a job into the network from outside, and its departure from the network. Applying Little's theorem to the entire network, we get

$$L = \gamma W , \qquad (4.9)$$

where $\gamma = \gamma_1 + \gamma_2 + \ldots + \gamma_N$ is the total external arrival rate. This, together with (4.8) and (4.6), yields (when all nodes are single-server ones)

$$W = \frac{1}{\gamma} \sum_{i=1}^{N} \frac{\rho_i}{1 - \rho_i} . \qquad (4.10)$$

One may also be interested in various other passage times associated with jobs in the network. Consider the average interval, R_i, between the arrival of a job at node i, and its subsequent departure from the network (this is a passage time from node i to the outside world). Because of the memoryless routeing, it does not matter whether the job arrived at node i from outside, or from another node, or from node i. First, the job has to wait and receive service at node i, which takes time W_i, on the average. After that, if it goes to node j (with probability q_{ij}), its average remaining time in the network will be R_j. Hence, we can write a set of linear equations

$$R_i = W_i + \sum_{j=1}^{N} q_{ij} R_j \; ; \; i = 1, 2, \ldots, N , \qquad (4.11)$$

or, in matrix and (column) vector form, $\mathbf{R} = \mathbf{W} + Q\mathbf{R}$. These equations determine the average passage times uniquely, since the matrix $I - Q$ is non-singular.

Another quantity of interest is the average number of visits, v_i, that a job makes to node i during its residence in the network. To find that number, note that an externally arriving job joins node i with probability γ_i/γ (this is the fraction of all external arrivals that come into node i). Thereafter, for every visit the job makes to node j, there will be a visit

4.1 Open networks

to node i with probability q_{ji}. This leads to the set of equations

$$v_i = \frac{\gamma_i}{\gamma} + \sum_{j=1}^{N} v_j q_{ji} \ ; \ \ i = 1, 2, \ldots, N. \tag{4.12}$$

Comparing (4.12) with the traffic equations, (4.2), we note that the coefficient matrices are identical, while the free terms differ by a constant factor, $1/\gamma$. We conclude, therefore, that the solutions also differ by the same constant factor:

$$v_i = \frac{\lambda_i}{\gamma}. \tag{4.13}$$

(A different argument in favour of (4.13) is the following: an average of γ jobs enter the network per unit time; each of those jobs makes an average of v_i visits to node i; hence, the total average number of arrivals into node i per unit time is γv_i.)

Every time a job visits node i, it requires service b_i, on the average. Therefore, the total average service, s_i, that one job requires from node i during its life in the network is equal to

$$s_i = v_i b_i = \frac{\rho_i}{\gamma} \ ; \ \ i = 1, 2, \ldots, N. \tag{4.14}$$

Rewriting this as $\rho_i = \gamma s_i$, we see that the load ρ_i can be interpreted as the total average demand for node i service that enters the network per unit time.

The total average amount of service, s, that a job requires during its residence in the network is

$$s = \sum_{i=1}^{N} v_i b_i = \frac{1}{\gamma} \sum_{i=1}^{N} \rho_i. \tag{4.15}$$

This is a lower bound on the average response time, since the latter includes waiting in queues.

Example

3. In the four-node network of example 2 (figure 4.5), the average service times are $b_1 = b_2 = 1$, $b_3 = b_4 = 2$. What is the largest lower bound on the average response time, W? What restriction must be imposed on the external arrival rate, γ, in order to achieve $W < 20$?

From the solution of the traffic equations we found that the total arrival rates are $\lambda_i = 2.5\gamma$, $i = 1, 2, 3, 4$. The offered loads are therefore $\rho_1 = \rho_2 = 2.5\gamma$, $\rho_3 = \rho_4 = 5\gamma$. The condition for stability is $\gamma < 0.2$.

The total average number of jobs in the network is equal to

$$L = L_1 + L_2 + L_3 + L_4 = \frac{5\gamma}{1 - 2.5\gamma} + \frac{10\gamma}{1 - 5\gamma} .$$

The average response time is given by

$$W = \frac{L}{\gamma} = \frac{5}{1 - 2.5\gamma} + \frac{10}{1 - 5\gamma} .$$

This is an increasing function of γ and therefore the largest lower bound is obtained by setting $\gamma = 0$, which yields $W > 15$. Another way of determining that bound is to compute the total average service requirement, s: the average number of visits to every node is $\lambda_i/\gamma = 2.5$; hence, $s = 2.5 + 2.5 + 5 + 5 = 15$.

The requirement $W < 20$ leads to a quadratic inequality for γ:

$$50\gamma^2 - 20\gamma + 1 > 0 ,$$

which, together with the stability condition, implies

$$\gamma < \frac{2 - \sqrt{2}}{10} \approx 0.0586 .$$

* * *

When constructing a queueing network model, it is sometimes easier to estimate, or to make assumptions, about the total average service requirements, s_i, than about the requirements per visit, b_i, and the routeing probabilities, q_{ij}. Given the quantities s_i, and the external arrival rates, one can compute the loads ρ_i without having to solve the traffic equations. This provides the performance measures L_i, L and W. The total average time that a job spends at node i during its life in the network, D_i, can also be obtained:

$$D_i = v_i W_i = \frac{v_i b_i}{1 - \rho_i} = \frac{s_i}{1 - \rho_i} . \quad (4.16)$$

What cannot be derived without knowing b_i and q_{ij}, is the average time spent at node i per visit, W_i, and also the average passage time, R_i.

Remark. Jackson's theorem refers to the distribution of the network state, (4.5), seen by a random observer. According to the PASTA property (section 2.3.1), that is also the distribution seen by jobs arriving into the network from outside. What is the distribution seen by a job arriving into a node from another node? The answer to that qestion is not obvious, since the internal arrival processes are not, in general, Poisson. However, it can be shown that all jobs arriving into any node can be treated as random observers:

Theorem 4.2 *Let $\pi_i(n_1, n_2, \ldots, n_N)$ be the steady-state probability that there are n_1 jobs at node 1, n_2 jobs at node 2, ..., n_N jobs at node N, at the moment when a job is about to enter node i (that is, the job is deemed to have left its previous node, if any, and not to have joined node i). Then*

$$\pi_i(n_1, n_2, \ldots, n_N) = p(n_1, n_2, \ldots, n_N),$$

where $p(n_1, n_2, \ldots, n_N)$ is given by (4.5).

This is the open network version of a result known as the 'arrival theorem'. An outline of the proof is given in exercise 4.

The arrival theorem implies that node i is seen as an isolated M/M/1 queue by all incoming jobs. Therefore, the distribution function, $F_i(x)$, of the time jobs spend at that node per visit is equal to

$$F_i(x) = 1 - e^{-(1-\rho_i)x/b_i}. \tag{4.17}$$

Moreover, when a job moves from node i to node j, the numbers of jobs left behind at node i and found at node j are independent of each other. The same applies to the times the job spends at the two nodes. Thus, the distribution of the passage time through two adjacent nodes can be obtained as the convolution of two exponential distributions.

Unfortunately, the above observation does not generalize to paths consisting of more than two nodes. There may be dependencies between the times spent at distant nodes. Consider, for example, the three-node network illustrated in figure 4.6. After completing service at node 1, jobs go to node 2 with probability q and to node 3 with probability $1-q$. The (external) arrival rate at node 1 is γ. It is readily seen that the steady-state arrival rates at nodes 2 and 3 are $q\gamma$ and γ, respectively. For stability, those rates have to satisfy $\gamma < 1/b_1$, $q\gamma < 1/b_2$ and $\gamma < 1/b_3$. Assume that these inequalities hold and the network is in steady state.

If a job takes the path $1 \to 2 \to 3$, its passage time through the network is the sum of the times spent at the three nodes: $t = t_1 + t_2 + t_3$. Now, t_1 and t_2 are independent random variables, and so are t_2 and t_3. However, t_1 and t_3 are dependent! To see that, suppose that t_1 is very large, so that the job leaves a long queue at node 1. While it passes

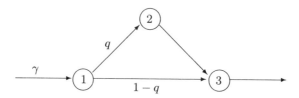

Fig. 4.6. A three-node network with dependent passage times

through node 2, many of the jobs from that long queue will overtake it by going directly to node 3, causing t_3 to be large.

To summarize, average passage times along arbitrary paths in a Jackson network are easy to obtain (the expectation of a sum is equal to the sum of expectations). On the other hand, the distributions of such passage times are intractable, except in some special cases.

Exercises

1. Consider an open Jackson network consisting of N nodes connected in a ring. The configuration is symmetric: after completing service at any node, jobs go to either of its two neighbours with probability 0.3 each way, and leave the network with probability 0.4. The external arrival rates at all nodes are equal to γ/N; the average service times at all nodes are equal to 1.

Find the total arrival rate into each node and the condition that γ must satisfy so that the network is stable. Determine the total average number of jobs in the network, the average response time, the total average required service time per job and the total average number of services that a job receives while in the network.

2. In a three-node Jackson network, the flow of jobs is governed by a routeing matrix, $Q = [q_{ij}]$, given by

$$Q = \begin{bmatrix} 0 & 0.2 & 0.8 \\ 0 & 0 & 0 \\ 1 & 0 & 0 \end{bmatrix}.$$

There are external arrivals at nodes 1 and 2, with rates 3 and 2, respectively. The average service times at nodes 1, 2 and 3 are 3, 10 and 4, respectively.

Find the average numbers of jobs at the three nodes and in the whole network. Also determine the average passage times, R_1 and R_2, from joining nodes 1 and 2, respectively, until leaving the network.

3. Let v_{ij} be the average number of visits that a job will make to node j before leaving the network, given that it is about to join node i. Show that these averages satisfy the following set of equations:

$$v_{ii} = 1 + \sum_{k=1}^{N} q_{ik} v_{kj} \; ; \; i = 1, 2, \ldots, N \; ,$$

$$v_{ij} = \sum_{k=1}^{N} q_{ik} v_{kj} \; ; \; i \neq j = 1, 2, \ldots, N \; .$$

Use those equations in the context of the network in the previous exercise and find the average numbers of visits to node 1 and to node 3 made by a job arriving into node 1. Compare the results with the unconditional averages v_1 and v_3, and explain the discrepancy.

4. Proof of arrival theorem. Use a vector, $\mathbf{n} = (n_1, n_2, \ldots, n_N)$, to denote the network state. Let $\mathbf{e}_i$ be the vector whose ith element is 1 and all others are 0. Thus, $\mathbf{n} + \mathbf{e}_i$ is the network state where there is one more job at node i than in state $\mathbf{n}$. Show that the steady-state distribution given by (4.5) satisfies

$$\lambda_i p(\mathbf{n}) = \frac{1}{b_i} p(\mathbf{n} + \mathbf{e}_i) \; ; \; i = 1, 2, \ldots, N \; .$$

Denote by $\pi_i(\mathbf{n})$ the probability that a job arriving into node i sees state $\mathbf{n}$, and by $\nu_i(\mathbf{n})$ the average number of arrivals into node i per unit time that see state $\mathbf{n}$. Since the total average number of arrivals into node i per unit time is λ_i, we can write

$$\pi_i(\mathbf{n}) = \frac{\nu_i(\mathbf{n})}{\lambda_i} \; ; \; i = 1, 2, \ldots, N \; .$$

Express $\nu_i(\mathbf{n})$ in terms of $p(\mathbf{n})$ by arguing as follows: a job arrives into node i and sees state $\mathbf{n}$ if (a) an external arrival into node i occurs when the state is $\mathbf{n}$, or (b) the state is $\mathbf{n} + \mathbf{e}_j$, a job completes service at node j and transfers to node i. The service completion rate at node j is $1/b_j$ (section 2.2). Hence,

$$\nu_i(\mathbf{n}) = \gamma_i p(\mathbf{n}) + \sum_{j=1}^{N} \frac{1}{b_j} p(\mathbf{n} + \mathbf{e}_j) q_{ji} \; .$$

Deduce from the above equations that $\pi_i(\mathbf{n}) = p(\mathbf{n})$.

4.2 Closed networks

There are many systems where jobs can be delayed at a number of nodes, and where the total number of jobs available, or admitted, is kept constant over long periods of time. Such systems are modelled by closed queueing networks, i.e. ones without external arrivals and departures. We have already encountered examples of that type in chapter 3. The finite-source model of section 3.1.1 (figure 3.3) can be regarded as a closed network with two nodes: one containing K user servers and one containing n processors. A fixed number, K, of jobs circulate among the two nodes at all times.

Here we shall consider closed queueing networks with N nodes, each of which contains either a single server, or an unlimited number of servers (at least as many as there are jobs). The latter are independent delay nodes; they can be used to model finite collections of users, as in figure 3.3, or may be introduced in order to model a non-zero transfer time between a pair of other nodes.

The reason for restricting the discussion to single-server and independent delay nodes is simplicity. Some generalizations will be mentioned later.

The average service time at node i is b_i ($i = 1, 2, \ldots, N$). The distribution of service times at a single-server node is assumed to be exponential, but at an independent delay node it may be general (at those nodes, we may talk about 'delay times' or 'think times', instead of service times). After completing service at node i, a job goes to node j with probability q_{ij} ($i, j = 1, 2, \ldots, N$). These routing probabilities now satisfy

$$\sum_{j=1}^{N} q_{ij} = 1 \; ; \; i = 1, 2, \ldots, N.$$

Thus, no job ever leaves the network. There are no external arrivals, either. The total number of jobs in the network is always constant, and is equal to the number that were there at time 0. Denote that number by K.

The state of the network at any time is described by the vector $(n_1, n_2, \ldots, n_N)$, where n_i is the number of jobs at node i. Since the only feasible states are those for which

$$\sum_{i=1}^{N} n_i = K, \qquad (4.18)$$

the state space is obviously finite. More precisely, the number of possible states, n, is equal to the number of ways in which the integer K can be partitioned into N non-negative components. That number is easily seen (see exercise 1) to be equal to

$$n = \binom{K+N-1}{N-1}. \qquad (4.19)$$

Denote by λ_i the total average number of jobs arriving into node i per unit time ($i = 1, 2, \ldots, N$). These jobs can only come from other nodes in the network. Following the same reasoning as in the case of an open network, we can write a set of traffic equations that the arrival rates must satisfy:

$$\lambda_i = \sum_{j=1}^{N} \lambda_j q_{ji} \; ; \; i = 1, 2, \ldots, N. \qquad (4.20)$$

These equations are now homogeneous, because of the absence of external arrival rates. Moreover, the coefficient matrix, $I - Q$, is singular, since all its row sums are 0. Hence, (4.20) has infinitely many solutions. Assuming that the rank of $I - Q$ is $N - 1$, all those solutions differ from each other by constant factors. If one of the λ_i values is fixed arbitrarily and the others are determined from the traffic equations, then the resulting quantities λ_1, λ_2, ..., λ_N are proportional to the true arrival rates. Similarly, the products $\rho_i = \lambda_i b_i$, obtained from that solution, are proportional to the true offered loads.

The steady-state distribution of the network state, $p(n_1, n_2, \ldots, n_N)$, is given by a result known as the 'Gordon–Newell theorem':

Theorem 4.3 *(W.J. Gordon and G.F. Newell) Let λ_1, λ_2, ..., λ_N be any solution of equations (4.20), and $\rho_i = \lambda_i b_i$, $i = 1, 2, \ldots, N$. Then*

$$p(n_1, n_2, \ldots, n_N) = \frac{1}{G} \prod_{i=1}^{N} \beta_i(n_i) \rho_i^{n_i}, \qquad (4.21)$$

where

$$\beta_i(n_i) = \begin{cases} 1 & \text{if node } i \text{ has a single server} \\ 1/n_i! & \text{if } i \text{ is an independent delay node} \end{cases}.$$

The 'normalizing' constant G is determined from the condition that the sum of all probabilities is 1:

$$G = \sum_{n_1+n_2+\ldots+n_N=K} \prod_{i=1}^{N} \beta_i(n_i) \rho_i^{n_i}. \qquad (4.22)$$

We omit the proof. Note that there are certain similarities between the Jackson and Gordon–Newell theorems. Both give the joint distribution as a product of factors corresponding to individual nodes. However, there are also some important differences. The ith factor in the right-hand side of (4.21) cannot be interpreted as the probability that there are n_i jobs at node i. Moreover, the numbers of jobs present at different nodes are not independent random variables (e.g., if there are K jobs at node 1, then all other nodes must be empty).

From a computational point of view, there is just one difficult step in implementing the solution suggested by the Gordon–Newell theorem, and hence obtaining various network performance measures. That is the summation which must be performed in order to evaluate the normalizing constant, G. The reason for the difficulty is of course the number of terms involved. The size of the state space, given by (4.19), increases rather quickly with N and K. Several algorithms exist for computing G efficiently (one of them is outlined in exercise 2). However, we shall not pursue that avenue any further because most of the performance measures of interest can be obtained in a simpler and more intuitively appealing way.

4.2.1 Mean value analysis

The approach that we are about to describe deals only with averages and is based on simple applications of Little's theorem; for that reason, it is called 'mean value analysis'. The quantities of interest are the average number of jobs at node i, L_i, and the average time a job spends at node i per visit, W_i. In addition, we shall wish to talk about the average time a job spends in the network, W, the average number of visits that a job makes to node i, v_i, and the throughput, T (i.e. the average number of jobs leaving the network per unit time). Obviously, in order to do this in the context of a closed network, it is necessary to re-introduce external arrivals and departures, without otherwise disturbing the model.

In a real system where the number of jobs is kept constant, the membership of the set of jobs usually changes. When a job is completed, it departs and is immediately replaced by a new one from outside. This is what will happen in our model. Assume that there is a special point on one of the arcs in the network, called the 'entry point' and denoted by 0. Suppose, without loss of generality, that the arc containing point 0 leads from node 1 to node 2 (it may also be a feedback arc from a node to itself). Whenever a job traverses that particular arc and passes

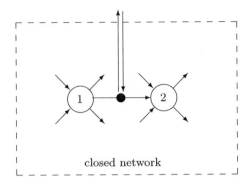

Fig. 4.7. Entry point on the arc from node 1 to node 2

through point 0, it changes into a new job. That is, the old job departs from the network, a new job enters it immediately and goes to node 2. The exchange is illustrated in figure 4.7.

Clearly, this modification of the model has no effect other than to transform the jobs from permanent entities into temporary ones. What was before the period between two consecutive passes of the same job through point 0 is now the residence time of a job in the network. Consequently, the performance measures W, v_i and T acquire meaning and interest.

The average numbers of visits, v_i, satisfy the set of equations (4.12). When the network is closed, the latter become

$$v_i = \sum_{j=1}^{N} v_j q_{ji} \ ; \ i = 1, 2, \ldots, N \ . \tag{4.23}$$

These equations do not have a unique solution. However, if one of the v_i s is known, all the others can be obtained from (4.23).

By the definition of the entry point, each job passes through it exactly once: $v_0 = 1$. This fact determines the average number of visits for the node that immediately precedes point 0. Indeed, if point 0 is on the arc from node 1 to node 2, as in figure 4.7, each visit to node 1 is followed by a passage through point 0 with probability q_{12}. Hence, $v_1 q_{12} = v_0 = 1$, or $v_1 = 1/q_{12}$.

Thus, fixing the position of the entry point enables us to solve equations (4.23) and find all average numbers of visits. Note that those quantities do not depend on the number of jobs in the network, K.

Since the average number of jobs entering the network per unit time is T, and each of them makes an average of v_i visits to node i, the total arrival rate into node i is

$$\lambda_i = Tv_i \; ; \; i = 1, 2, \ldots, N \, . \tag{4.24}$$

An application of Little's result to node i yields the familiar relation between the average number of jobs at that node, L_i, and the average time jobs spend there per visit, W_i:

$$L_i = \lambda_i W_i = Tv_i W_i \; ; \; i = 1, 2, \ldots, N \, . \tag{4.25}$$

Summing these over all nodes and remembering that the total number of jobs in the network is K, we get

$$K = T \sum_{i=1}^{N} v_i W_i \, ,$$

or

$$T = \frac{K}{\sum_{i=1}^{N} v_i W_i} \, . \tag{4.26}$$

The average time a job spends at node i, W_i, depends on the type of the node and on the number of jobs found there by an incoming job. If node i is an independent delay node, then there is no queue and W_i is equal to the average service time, b_i. If node i is a single-server node, and if when a job gets there it finds an average of Y_i jobs already present, then W_i is equal to Y_i+1 average service times. Introducing an indicator, δ_i, which has value 0 if node i is an independent delay node and 1 if it is a single-server node, we can write

$$W_i = b_i(1 + \delta_i Y_i) \; ; \; i = 1, 2, \ldots, N \, . \tag{4.27}$$

Remark. The network state, and in particular the state of node i, seen by an incoming job, has a different distribution from the state seen by a random observer. For instance, an incoming job can never see all K jobs at node i, because it itself cannot be among the jobs already present. That is why Y_i is different from L_i. Yet if one could somehow relate these two types of averages, all performance measures would be determined from equations (4.25), (4.26) and (4.27).

A way out of this difficulty is provided by a closed network version of the arrival theorem of the previous section. To state the result, it is convenient to describe the network state by a pair $(\mathbf{n}, K)$, where $\mathbf{n} = (n_1, n_2, \ldots, n_N)$ specifies the numbers of jobs present at nodes

4.2 Closed networks

$1, 2, \ldots, N$, and K is the total number of jobs in the network. If a random observer sees state $(\mathbf{n}, K)$, then the sum of the elements of $\mathbf{n}$ is K. On the other hand, if that state is seen by a job about to arrive into a node, then the elements of $\mathbf{n}$ add up to $K - 1$ because the observer is not counted as being present in any node.

The steady-state, or random observer distribution, $p(\mathbf{n}, K)$, given by the Gordon-Newell theorem, and the distribution seen by arriving jobs, are related as follows:

Theorem 4.4 *The probability, $\pi_i(\mathbf{n}, K)$, that a job which is about to arrive into node i sees state $(\mathbf{n}, K)$, is equal to*

$$\pi_i(\mathbf{n}, K) = p(\mathbf{n}, K - 1) \;.$$

This arrival theorem allows us to treat a job in transit from one node to another as a random observer. However, the network that is being observed contains $K - 1$ jobs instead of K. The claim does not lack intuition, since a job in transit cannot see itself at any node. The formal proof is non-trivial and will be omitted.

We can now develop a solution procedure based on recurrence relations with respect to the job population size, K. The quantities T, W_i, L_i and Y_i will be regarded as functions of K. The arrival theorem implies that

$$Y_i(K) = L_i(K - 1) \;\; ; \;\; i = 1, 2, \ldots, N \;. \tag{4.28}$$

This allows us to rewrite equations (4.27), (4.26) and (4.25) in the form

$$W_i(K) = b_i[1 + \delta_i L_i(K - 1)] \;\; ; \;\; i = 1, 2, \ldots, N \;. \tag{4.29}$$

$$T(K) = \frac{K}{\sum_{i=1}^{N} v_i W_i(K)} \;\; ; \;\; i = 1, 2, \ldots, N \;. \tag{4.30}$$

$$L_i(K) = v_i T(K) W_i(K) \;\; ; \;\; i = 1, 2, \ldots, N \;. \tag{4.31}$$

The obvious starting condition for an iterative procedure based on these recurrences is $L_i(0) = 0$ ($i = 1, 2, \ldots, N$). The first application of the procedure gives $W_i(1)$, $T(1)$ and $L_i(1)$; the second gives $W_i(2)$, $T(2)$ and $L_i(2)$, and so on, until the desired population size is reached.

The total average time that a job spends at node i during its life in the network, $D_i(K)$, is equal to

$$D_i(K) = v_i W_i(K) \;\; ; \;\; i = 1, 2, \ldots, N \;. \tag{4.32}$$

Similarly, the total average service time that a job requires from node i, s_i (that quantity does not depend on K), is equal to

$$s_i = v_i b_i \; ; \; i = 1, 2, \ldots, N \, . \qquad (4.33)$$

Sometimes the total average service requirements s_i can be estimated quite easily from measurements. They may therefore be given as the network parameters, instead of the individual average service times b_i and the routeing matrix Q. In that case, the recurrence schema (4.29), (4.30) and (4.31) is replaced by a simpler one which does not involve the visit averages v_i. The role of $W_i(K)$ is played by $D_i(K)$ and that of b_i by s_i:

$$D_i(K) = s_i[1 + \delta_i L_i(K-1)] \; ; \; i = 1, 2, \ldots, N \, . \qquad (4.34)$$

$$T(K) = \frac{K}{\sum_{i=1}^{N} D_i(K)} \; ; \; i = 1, 2, \ldots, N \, . \qquad (4.35)$$

$$L_i(K) = T(K) D_i(K) \; ; \; i = 1, 2, \ldots, N \, . \qquad (4.36)$$

The total average time that a job spends in the network, $W(K)$, is given by

$$W(K) = \sum_{i=1}^{N} v_i W_i(K) = \sum_{i=1}^{N} D_i(K) = \frac{K}{T(K)} \, . \qquad (4.37)$$

The last equality follows also from Little's result, applied to the entire network.

If node i contains a single server, its utilization, $U_i(K)$, is equal to the offered load:

$$U_i(K) = \lambda_i b_i = T(K) v_i b_i = T(K) s_i \, . \qquad (4.38)$$

This result provides us with a simple upper bound on the network throughput. Since the utilization of a server cannot exceed 1, we must have

$$T(K) \leq \min_i \frac{1}{s_i} \, , \qquad (4.39)$$

where the minimum is taken over all single-server nodes. The node for which that minimum is reached, i.e. the one with the largest total service requirement, is called the 'bottleneck node'.

The solution procedure that we have described evaluates $2N+1$ quantities (W_i, L_i and T, or D_i, L_i and T), for each population size up to the desired one. The necessity of stepping through all the intermediate

4.2 Closed networks

values of K can be avoided if one is willing to accept an approximate solution instead of an exact one. The idea is to relate the job-observed and random-observed node averages, Y_i and L_i, at the same population level. One such relation that has been suggested is

$$Y_i(K) = \frac{K-1}{K} L_i(K) \ ; \ i = 1, 2, \ldots, N \ . \tag{4.40}$$

This expression is of course exact for $K = 1$. It is also asymptotically exact for large values of K. In general, experience indicates that the approximations obtained with the aid of (4.40) are reasonably accurate for most practical purposes.

The use of (4.40) reduces the relations (4.29), (4.30) and (4.31), or (4.34), (4.35) and (4.36), to fixed-point equations expressing the $2N+1$ unknowns in terms of themselves. All but one of the unknowns can be eliminated explicitly. For instance, replacing $L_i(K-1)$ in (4.34) with the right-hand side of (4.40) and using (4.36), we get

$$D_i(K) = \frac{s_i K}{K - (K-1)\delta_i s_i T(K)} \ ; \ i = 1, 2, \ldots, N \ .$$

This allows $D_i(K)$ to be eliminated from (4.35), turning the latter into a fixed-point equation for the throughput alone, of the form

$$T = f(T) \ .$$

Such equations are usually solved by iteration: one starts with an initial approximation, T_0 (e.g., $T_0 = 1/s_i$, for some i); from then on, iteration n computes

$$T_n = f(T_{n-1}) \ ,$$

until the difference between successive values becomes sufficiently small. Those iterations may be faster than the ones solving the exact recurrences.

Example

1. Window flow control. Consider the closed network illustrated in figure 4.8. Messages originating at node 1 are sent to node N and have to pass through intermediary nodes 2, 3, ..., $N-1$. Each of these transfers is successful with probability α and fails with probability $1 - \alpha$; in the latter case, the message returns to node 1 and has to be retransmitted. After node N, the message departs and is replaced by a new one at node 1 (i.e. the entry point is on the arc from node N to node 1). The total

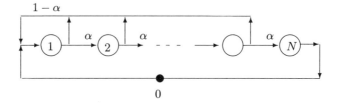

Fig. 4.8. Nodes in sequence with a window of size K

number of messages in the network at any time is K: this is called the 'window size'. The average service times at all nodes are equal to b.

The visit numbers, v_i, satisfy the following equations:

$$v_i = \alpha v_{i-1} \; ; \; i = 2, 3, \ldots, N-1$$

(the equation for v_1 is more complicated, but it is redundant). In addition, since jobs leave the network after node N, we have $v_N = 1$. Therefore,

$$v_i = \frac{1}{\alpha^{N-i}} \; ; \; i = 1, 2, \ldots, N \, .$$

The total required service times from the different nodes are

$$s_i = \frac{b}{\alpha^{N-i}} \; ; \; i = 1, 2, \ldots, N \, .$$

The performance measures (of which the main one is $T(K)$) can now be obtained by means of the recurrences (4.34), (4.35) and (4.36). Clearly, the bottleneck is node 1; the throughput cannot exceed α^{N-1}/b.

Exercises

1. The state of a closed network with N nodes and K jobs (or a partition of the integer K into N components) can be represented by a string of N zeros and K ones. Each node, starting with the first, contributes to the string a 0 followed by as many 1s as there are jobs present at that node. Thus, for a network with three nodes and five jobs, the string 01001111 represents the state where there is one job at node 1, no jobs at node 2 and four jobs at node 3.

Establish (4.19) by arguing that the number of possible states is equal to the number of ways of choosing the positions of $N-1$ zeros among $K+N-1$ digits.

2. The normalizing constant G which appears in the Gordon–Newell theorem depends on the number of nodes and on the number of jobs. To make that dependence explicit, it is commonly denoted by $G_N(K)$.

Suppose that all nodes in the network are single-server ones. Then expression (4.22) becomes

$$G_N(K) = \sum_{n_1+n_2+\ldots+n_N=K} \prod_{i=1}^{N} \rho_i^{n_i}.$$

Split the right-hand side into two sums: the first extends over those states for which $n_N = 0$, and the second over the states for which $n_N > 0$. Demonstrate that the resulting expression can be written as

$$G_N(K) = G_{N-1}(K) + \rho_N G_N(K-1).$$

This recurrence suggests an efficient algorithm for computing $G_N(K)$, starting with the initial conditions $G_0(K) = 0$ ($K = 1, 2, \ldots$) and $G_N(0) = 1$ ($N = 1, 2, \ldots$).

3. Consider a closed network consisting of three single-server nodes. After leaving node 1, a job goes to node 2 with probability 0.4, and to node 3 with probability 0.6. After node 2, jobs always go to node 3. After node 3, a job goes to node 1 with probability 0.5, and leaves the network with probability 0.5; in the latter case, a new job immediately enters node 2 (the entry point is on an arc from node 3 to node 2). At time 0, there are two jobs in the network. The average service times at nodes 1, 2 and 3 are 2, 5 and 0.5, respectively.

Find the steady-state throughput and average response time in this network. Identify the bottleneck node and hence give an upper bound for the throughput that may be achieved by introducing more jobs at time 0.

4. A multiprogrammed computer under heavy demand, and also some organizations dealing with customers, can be modelled by a closed network of the type illustrated in figure 4.9. Node 1 represents a central server (a processor or a reception desk), while nodes $2, 3, \ldots, N$ are various peripheral servers. After visiting node 1, a job goes to node i with probability α_i ($i = 2, 3, \ldots, N$), and leaves the network with probability $\alpha_1 = 1 - \alpha_2 - \ldots - \alpha_N$; in the latter case, a new job is immediately introduced into node 1. Thus, the entry point is on the feedback arc from node 1 to node 1. After a visit to node i ($i > 1$), jobs go back to

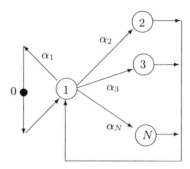

Fig. 4.9. A closed network with central and peripheral nodes

node 1. Node i has a single exponential server with mean service time b_i ($i = 1, 2, \ldots, N$). The number of jobs allowed in the network is K.

Find the average number of visits a job makes to node i. Implement the recurrence relations that determine the performance measures. In the special case when $b_i = b$ and $\alpha_i = 1/N$ ($i = 1, 2, \ldots, N$), solve those recurrences explicitly and write closed-form expressions for $T(K)$ and $W(K)$.

5. In a fully connected closed network with N nodes and K jobs, all elements of the routeing matrix are equal: $q_{ij} = 1/N$ ($i, j = 1, 2, \ldots, N$). The average service times at all nodes are equal to b.

Place the entry point on any arc (e.g. on the feedback arc from node 1 to node 1). Using the symmetry of the model, derive closed-form expressions for all average performance measures.

4.3 Multiclass networks

The restriction that all jobs are statistically identical can be relaxed considerably, in both open and closed networks. Here we shall assume that there are C different job types, and that the paths followed by type c jobs are governed by a separate routeing matrix, Q_c ($c = 1, 2, \ldots, C$). The (i, j)th element of that matrix, q_{cij}, gives the probability that a job of type c, having completed service at node i, goes to node j. The network can now be open with respect to all job types, closed with respect to all job types, or mixed, i.e. open for some types and closed for others (the definitions of 'open' and 'closed' are the same as before).

Having admitted the possibility of different routeing patterns for the

4.3 Multiclass networks

different job types, it would also be nice to allow arbitrary service time distributions and different scheduling policies at different nodes. Unfortunately, one cannot go too far in that direction and yet remain within the realm of solvable models. The following constraints have to be imposed on the nodes where queueing takes place:

1. If the scheduling strategy at node i is FIFO, then the required service times of all jobs (on each visit to that node) must be exponentially distributed, with the same mean, b_i.
2. If the required service times at node i have different distribution functions for the different job types—say $F_{ci}(x)$ for type c—then the scheduling strategy there must be 'symmetric', as defined in section 3.3.2.

At independent delay nodes, where there is no queueing, the service time distributions may be arbitrary and different for the different job types.

Examples of symmetric scheduling policies are Processor-Sharing and Last-In-First-Out Preemptive-Resume (see section 3.3).

Queueing networks with multiple job types, where all nodes are either independent delay nodes or single-server ones satisfying conditions 1 and 2, have product-form solutions and are amenable to mean value analysis. The procedures for obtaining performance measures are similar to the ones described in the previous section, so our presentation here will be less detailed. All networks are assumed to be in steady state.

Consider first the open case. Jobs of type c arrive externally into node i in a Poisson stream with rate γ_{ci} ($c = 1, 2, \ldots, C$; $i = 1, 2, \ldots, N$). For each c, at least one of the rates γ_{ci} is non-zero. The total rate at which type c jobs arrive into node i, λ_{ci}, can be determined by solving the following set of traffic equations.

$$\lambda_{ci} = \gamma_{ci} + \sum_{j=1}^{N} \lambda_{cj} q_{cji} \ ; \ c = 1, 2, \ldots, C \ ; \ i = 1, 2, \ldots, N \ . \quad (4.41)$$

In fact, what we have here is C independent sets of equations: one for each job type. More generally, jobs may be allowed to change type as they move from node to node: the routeing may be governed by probabilities q_{cdij}, that a job of type c leaving node i will go to node j as a job of type d. We shall not consider that generalization.

Having determined λ_{ci} from (4.41), the load of type c at node i is equal to $\rho_{ci} = \lambda_{ci} b_{ci}$ (of course, if node i is a single-server FIFO node then

$b_{ci} = b_i$ for all c). The total load at node i is $\rho_i = \rho_{1i} + \rho_{2i} + \cdots + \rho_{Ci}$. In order that the network may be stable, the inequalities $\rho_i < 1$ must be satisfied for all single-server nodes.

Node i can be treated as an isolated and independent queueing system subjected to the loads ρ_{ri} ($r = 1, 2, \ldots, R$). If it is an independent delay node, i.e. if there are infinitely many servers available, then the average number of type c jobs present, L_{ci}, and the average time type c jobs spend in it (per visit), W_{ci}, are equal to $L_{ci} = \rho_{ci}$ and $W_{ci} = b_{ci}$, respectively. In the case of a single-server node, the corresponding expressions are

$$L_{ci} = \frac{\rho_{ci}}{1 - \rho_i} \quad ; \quad c = 1, 2, \ldots, C. \tag{4.42}$$

$$W_{ci} = \frac{b_{ci}}{1 - \rho_i} \quad ; \quad c = 1, 2, \ldots, C. \tag{4.43}$$

These results are valid for the FIFO scheduling policy, provided that all service times are exponentially distributed with the same mean. They are also valid for any symmetric policy, with arbitrary distributions of the required service times (see section 3.3.2).

The average number of visits that a job of type c makes to node i, v_{ci}, is given by an expression analogous to (4.13):

$$v_{ci} = \frac{\lambda_{ci}}{\gamma_c} \quad ; \quad c = 1, 2, \ldots, C, \tag{4.44}$$

where $\gamma_c = \gamma_{c1} + \gamma_{c2} + \ldots + \gamma_{cN}$ is the total external arrival rate for type c (the type c throughput). The total average amount of service that a job of type c requires from node i is $s_{ci} = v_{ci} b_{ci} = \rho_{ci}/\gamma_c$.

The total average time that a type c job spends at node i during its life in the network, D_{ci}, is given by

$$D_{ci} = v_{ci} W_{ci} = \begin{cases} s_{ci} & \text{if an independent delay node} \\ s_{ci}/(1 - \rho_i) & \text{if a single-server node} \end{cases}. \tag{4.45}$$

The total average response time for type c, W_c, is equal to

$$W_c = \sum_{i=1}^{N} D_{ci} \quad ; \quad c = 1, 2, \ldots, C. \tag{4.46}$$

Average passage times conditioned upon the initial node can be obtained as was done in the derivation of (4.11).

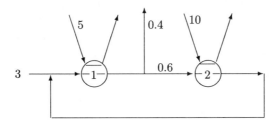

Fig. 4.10. An open network with three job types

Example

1. Two computer sites are connected by a communication channel which allows jobs to be transferred from one to the other. There are three job types, with the following behaviour:

Type 1. External Poisson arrivals into node 1 with rate 5 jobs per minute. The average required service time is equal to 4 seconds. After completion of service at node 1, these jobs leave the network.

Type 2. External Poisson arrivals into node 2 with rate 10 jobs per minute. Average required service time is 3 seconds. These jobs depart from the network after completing at node 2.

Type 3. External Poisson arrivals into node 1 with rate 3 jobs per minute. Service at that node is followed by a transfer to node 2, with probability 0.6, and by a departure from the network with probability 0.4. After node 2, these jobs return to node 1. Their average required service times at nodes 1 and 2 are 4 and 6 seconds, respectively.

The flow of jobs in this network is illustrated in figure 4.10.

Assume that at both nodes there is a single server using a Processor-Sharing scheduling policy. The performance measures of interest are the average response times for the three job types, W_1, W_2 and W_3.

Obviously, type 1 contributes only to the load at node 1. Similarly, type 2 contributes only to the load at node 2. Those contributions are $\rho_{11} = 5 \times (4/60) = 1/3$ and $\rho_{22} = 10 \times (3/60) = 0.5$, respectively. Jobs of type 3 circulate between the two nodes. The traffic equations for type 3 are

$$\lambda_{31} = 3 + \lambda_{32},$$
$$\lambda_{32} = 0.6\lambda_{31}.$$

These yield $\lambda_{31} = 7.5$ and $\lambda_{32} = 4.5$. Hence, the type 3 loads at the two nodes are $\rho_{31} = 7.5 \times (4/60) = 0.5$ and $\rho_{32} = 4.5 \times (6/60) = 0.45$, respectively. Steady state exists because the total loads at nodes 1 and 2, $\rho_1 = 5/6$ and $\rho_2 = 0.95$, are both less than 1.

The average response times for types 1 and 2 are obtained immediately, since those jobs visit their respective nodes only once. Applying (4.43), we get

$$W_1 = \frac{4}{50-60} = 0.4 \ ; \ W_2 = \frac{3}{60-57} = 1 \ .$$

In order to determine W_3, we need the average number of visits, v_{31} and v_{32}, that type 3 jobs make to nodes 1 and 2. These are given by (4.44):

$$v_{31} = \frac{7.5}{3} = 2.5 \ ; \ v_{32} = \frac{4.5}{3} = 1.5 \ .$$

Next, (4.45) yields the total average times that type 3 jobs spend at nodes 1 and 2:

$$D_{31} = \frac{10}{60-50} = 1 \ ; \ D_{32} = \frac{9}{60-57} = 3 \ .$$

Consequently, the average response time for jobs of type 3 is equal to $W_3 = 4$ minutes.

4.3.1 Closed multiclass networks

Suppose now that all external arrival rates are 0 and all the row-sums of all routeing matrices are equal to 1. The job population in the network is fixed, and is composed of K_c jobs of type c, $c = 1, 2, \ldots, C$. Jobs leave the network and are immediately replaced by new ones, using entry points which may be different for the different types. All nodes are either independent delay nodes or single-server ones satisfying conditions 1 and 2 of the previous subsection.

The traffic equations for the closed network are

$$\lambda_{ci} = \sum_{j=1}^{N} \lambda_{cj} q_{cji} \ ; \ c = 1, 2, \ldots, C \ ; \ i = 1, 2, \ldots, N \ . \quad (4.47)$$

Being homogeneous, these equations do not determine the arrival rates uniquely. They do so only up to C multiplicative constants: one for each job type. The average numbers of visits that jobs of type c make to the various nodes, v_{ci}, also satisfy equations (4.47). However, these last averages can be determined uniquely, because one of them is known, for

4.3 Multiclass networks

each c. For example, if the type c entry point is on the arc from node 1 to node 2, then v_{c1} is given by

$$v_{c1} = \frac{1}{q_{c12}}$$

(the argument from section 4.2.1 applies without change).

Having obtained the average numbers of visits, v_{ci}, we can write recurrence mean value equations for the type c throughput, T_c, average number of jobs at node i, L_{ci}, and average time spent on each visit to node i, W_{ci}. These quantities depend on the population size vector, $\mathbf{K} = (K_1, K_2, \ldots, K_R)$. The result that enables the recurrence equations to be written is a relation between the number of type r jobs seen at node i by an incoming type c job, $Y_{ri}^{(c)}$, and the similar number seen by a random observer, L_{ri}. That relation, provided by yet another version of the arrival theorem, can be stated in a form analogous to (4.28):

$$Y_{ri}^{(c)}(\mathbf{K}) = L_{ri}(\mathbf{K} - \mathbf{e}_c) , \tag{4.48}$$

where $\mathbf{e}_c$ is a C-dimensional vector whose cth element is 1 and all other elements are 0. In other words, a type c job coming into node i sees a state distribution that a random observer would see if there were one less type c job in the network.

The recurrence equations resulting from (4.48) have almost the same form as those in section 4.2.1. For $c = 1, 2, \ldots, C$ we have

$$W_{ci}(\mathbf{K}) = b_{ci}\left[1 + \delta_i \sum_{r=1}^{C} L_{ri}(\mathbf{K} - \mathbf{e}_c)\right] \;\; ; \;\; i = 1, 2, \ldots, N . \tag{4.49}$$

$$T_c(\mathbf{K}) = \frac{K_c}{\sum_{i=1}^{N} v_{ci} W_{ci}(\mathbf{K})} . \tag{4.50}$$

$$L_{ci}(\mathbf{K}) = T_c(\mathbf{K}) v_{ci} W_{ci}(\mathbf{K}) \;\; ; \;\; i = 1, 2, \ldots, N . \tag{4.51}$$

The definition of δ_i is the same as before. The fact that (4.49) holds for single-server nodes with different service requirements is a consequence of the properties of the symmetric scheduling policies.

The initial conditions are $L_{ci}(\mathbf{0}) = 0$ ($c = 1, 2, \ldots, C$; $i = 1, 2, \ldots, N$). Note that for each population vector there are $(2N+1)C$ equations. Also, in order to reach the desired population sizes, the procedure needs to step through a total of $(K_1 + 1)(K_2 + 1)\ldots(K_R + 1)$ intermediate population vectors. Clearly, when there are many job types and large populations for each type, the computational task is formidable.

As in the single-type case, adequate approximations can be obtained with much less effort by turning the recurrence equations into fixed-point ones. This is achieved by relating the averages associated with the population vector $\mathbf{K} - \mathbf{e}_c$ to those associated with the vector $\mathbf{K}$. One such relation which gives reasonably accurate results is the following:

$$L_{ri}(\mathbf{K} - \mathbf{e}_c) = \begin{cases} L_{ri}(\mathbf{K}) & \text{if } r \neq c \\ (K_c - 1)L_{ri}(\mathbf{K})/K_c & \text{if } r = c \,. \end{cases} \quad (4.52)$$

The total average time that a type c job spends at node i, $D_{ci}(\mathbf{K})$, and the total average time that it spends in the network, $W_c(\mathbf{K})$, are obtained from

$$D_{ci}(\mathbf{K}) = v_{ci} W_{ci}(\mathbf{K}) \,, \quad (4.53)$$

and

$$W_c(\mathbf{K}) = \sum_{i=1}^{N} D_{ci}(\mathbf{K}) = \frac{K_c}{T_c(\mathbf{K})} \,. \quad (4.54)$$

If, instead of being given the routeing matrices as network parameters, one has the total average service requirements for type c at node i, s_{ci}, then equations (4.49), (4.50) and (4.51) become slightly simpler:

$$D_{ci}(\mathbf{K}) = s_{ci}\left[1 + \delta_i \sum_{r=1}^{C} L_{ri}(\mathbf{K} - \mathbf{e}_c)\right] \;;\; i = 1, 2, \ldots, N \,. \quad (4.55)$$

$$T_c(\mathbf{K}) = \frac{K_c}{\sum_{i=1}^{N} D_{ci}(\mathbf{K})} \,. \quad (4.56)$$

$$L_{ci}(\mathbf{K}) = T_c(\mathbf{K}) D_{ci}(\mathbf{K}) \;;\; i = 1, 2, \ldots, N \,. \quad (4.57)$$

If the network is mixed, then the vector $\mathbf{K}$ describes the population sizes of the closed job types only. The recurrences are with respect to those population sizes. The state of an open job type, r, seen at node i by an incoming job of type c (regardless of whether the latter belongs to an open or closed type), has the same distribution as the corresponding state seen by a random observer. In particular, when r is an open job type,

$$Y_{ri}^{(c)}(\mathbf{K}) = L_{ri}(\mathbf{K}) \;;\; i = 1, 2, \ldots, N \,.$$

Finally, let us mention some further generalizations that are possible in queueing network models, but have not been covered here. First, the service at a node may be provided by a server whose speed depends on the number of jobs present (but not on their types). The nodes with a single server of constant speed, those with a finite number of parallel servers and those with infinitely many parallel servers can be considered as special cases of a node with a state-dependent server.

We have mentioned that the type of a job may change as it follows its path through the network. By using such changes, and by introducing artificial job types, one could extend considerably the way the routeing of jobs is described. Already, by defining types appropriately, it is possible to model jobs that take different fixed paths (rather than probabilistic ones) through the network. If type changes are allowed, one could model both random and deterministic routeing behaviour where the future path taken by a job depends on its past.

A more recent development involves the introduction of 'negative customers'. When a negative job arrives into a node, it causes one of the ordinary ones there to disappear, or perhaps to move to another node without receiving service. The main motivation for these special jobs was to model inhibiting impulses in neural networks.

The price of adopting any of these generalizations is that the mean value analysis no longer applies. The product form of the joint network state distribution has to be used, with all the attending difficulties of computing normalization constants (in the case of closed networks).

Exercises

1. In the three-node network illustrated in figure 4.6, there are two types of jobs, arriving externally into node 1 at rates γ_1 and γ_2, respectively. Type 1 jobs follow the path $1 \to 2 \to 3$; their average service times at the three nodes are b_{11}, b_{12} and b_{13}, respectively. Type 2 jobs follow the path $1 \to 3$ and have average service times b_{21} and b_{23}, respectively. The scheduling policy at nodes 1 and 3 is Processor-Sharing.

Write expressions for the performance meaures of type 1 and type 2 jobs.

2. A finite-source system (see section 3.1.1) consists of K users sending jobs to a single server. The users are of two types: K_1 of them have average think times τ_1 and average service times b_1; the corresponding parameters for the other $K_2 = K - K_1$ users are τ_2 and b_2, respectively. This system can be modelled by the two-node closed network in figure 4.11, where node 1 is an independent delay node and node 2 contains the single server and its queue. Assume that the scheduling policy at node 2 is symmetric.

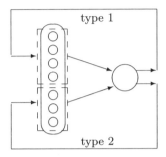

Fig. 4.11. A finite-source system with two job types

Implement the solution of the recurrence relations and find the average response times for the type 1 and type 2 users.

3. Turn the example in this section (figure 4.10) into a closed network by replacing the external arrivals and departures with feedback loops. Solve the resulting model for different population vectors (K_1, K_2, K_3).

4.4 Literature

The original works by Jackson [5] and Gordon and Newell [4] were motivated principally by problems in manufacturing, where components go through several stages of operation in the course of a production line. The first results on networks with multiple job types were obtained by Baskett et al. [1], and Kelly [6,7]. The fact that all symmetric scheduling policies have product-form solutions was discovered by Kelly [7].

Different versions of the arrival theorem were established independently by Kelly [7], Sevcik and Mitrani [14], Lavenberg and Reiser [9] and Melamed [11]. The mean value approach to the solution of closed networks was proposed by Reiser [12]; an extensive presentation of that method, including applications and approximations, can be found in Lazowska et al. [10]. A thorough mathematical treatment of queueing networks is provided in Walrand [15]. The numerical implementation of product-form solutions for closed networks is considered in Bruell and Balbo [2].

Other books containing material on queueing networks are Lavenberg [8] and Sauer and Chandy [13]. Networks with positive and negative customers were first studied by Gelenbe [3].

References

1. F. Baskett, K.M. Chandy, R.R. Muntz and F.G. Palacios, "Open, Closed and Mixed Networks of Queues with Different Classes of Customers", *JACM*, **22**, 248–260, 1975.
2. S.C. Bruell and G. Balbo, *Computational Algorithms for Closed Queueing Networks*, North-Holland, 1980.
3. E. Gelenbe, "Random Neural Networks with Negative and Positive Signals and Product Form Solution", *Neural Computation*, **1**, 502–510, 1989.
4. W.J. Gordon and G.F. Newell, "Closed Queueing Systems With Exponential Servers", *Operations Res.*, **15**, 266–278, 1967.
5. J.R. Jackson, "Networks of Waiting Lines", *Operations Res.*, **5**, 518–521, 1957.
6. F.P. Kelly, "Networks of Queues", *Adv. Appl. Prob.*, **8**, 416–432, 1976.
7. F.P. Kelly, *Reversibility and Stochastic Networks*, John Wiley, 1979.
8. S.S. Lavenberg, *Computer Performance Modelling Handbook*, Academic Press, 1982.
9. S.S. Lavenberg and M. Reiser, "Stationary State Probabilities at Arrival Instants for Closed Queueing Networks with Multiple Types of Customers", *J. Appl. Prob.*, **17**, 1048–1061, 1980.
10. E.D. Lazowska, J. Zahorjan, G.S. Graham and K.C. Sevcik, *Quantitative System Performance*, Prentice-Hall, 1984.
11. B. Melamed, "On Markov Jump Processes Imbedded in Jump Epochs and Their Queueing-Theoretic Applications", *Math. Oper. Res.*, **7**, 111–128, 1982.
12. M. Reiser, "Mean Value Analysis of Queueing Networks", in *Procs., 4th Int. Symp. on Modelling and Perf. Eval.* (edited by M. Arato, E. Gelenbe and A. Butrimenko), North-Holland, 1979.
13. C.H. Sauer and K.M. Chandy, *Computer Systems Performance Modelling*, Prentice-Hall, 1981.
14. K.C. Sevcik and I. Mitrani, "The Distribution of Queueing Network States at Input and Output Instants", *JACM*, **28**, 358–371, 1981.
15. J. Walrand, *An Introduction to Queueing Networks*, Prentice-Hall, 1988.

5

Markov chains and processes

A 'random process', also called a 'stochastic process', is defined formally as a set of random variables, $\{X_t\}$, indexed by a parameter t. The latter is usually interpreted as time: X_t represents the state of some system at time t. The following are examples of stochastic processes:

1. The number of messages sent across a communication channel during the interval $(0, t)$.
2. The size of a colony of bacteria, t hours after planting.
3. The number of patients in a dentist's waiting room at time t of a working day.
4. The numbers of jobs in the N nodes of a queueing network, $(n_1, n_2, \ldots, n_N)$, at time t of its operation (this process is vector-valued, or multi-dimensional).

In general, the future behaviour of a system depends on its past. That is, the states X_{t1} and X_{t2} at two different moments in time are dependent random variables. The challenge of the mathematical modelling of processes is to account for these dependencies as accurately as possible, while keeping the model tractable.

A class of stochastic processes exhibiting a limited, but very useful, form of state dependency was defined by A.A. Markov. These processes satisfy the so-called 'Markov property'. To illustrate the definition, consider figure 5.1, where a process is observed at a given moment, t. The current state of the process is X_t; its past history consists of the states X_u, for $u < t$; at time y in the future, the state will be X_y.

The Markov property. Given the the current state, X_t, the distribution of any future state, X_y, does not depend on the past history, $\{X_u : u < t\}$.

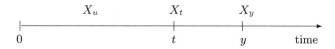

Fig. 5.1. Past, present and future

In other words, the future depends on the past only through its dependency on the present. In order to predict (probabilistically) the future behaviour of the process, it is sufficient to know the current state; there is no need to keep information about past states.

Processes which have the Markov property are 'Markov processes'. They are among the most important tools of probabilistic modelling.

The set of values taken by the random variables X_t is called the 'state space' of the process. We shall be interested in Markov processes whose state spaces are discrete. That is, the states can be enumerated using non-negative integers, $0, 1, \ldots$. If, in addition, the time parameter is also discrete, i.e. if the process is observed at time instants $0, 1, \ldots$, then the process is referred to as a 'Markov chain'. By implication, the term 'Markov process' is reserved for the case where the time parameter is continuous.

5.1 Markov chains

Let $X = \{X_n : n = 0, 1, \ldots\}$ be a Markov chain whose state space, $\{0, 1, \ldots\}$, may be finite or infinite. The Markov property says, in this case, that given the state of X at time n, its state at time $n + 1$ is independent of the states at times $0, 1, \ldots, n - 1$:

$$P(X_{n+1} = j \mid X_0, X_1, \ldots, X_n) = P(X_{n+1} = j \mid X_n). \qquad (5.1)$$

Hence, the evolution of the Markov chain is completely described by the 'one-step transition probabilities', $q_{i,j}(n)$, that the chain will move to state j at time $n + 1$, given that it is in state i at time n:

$$q_{i,j}(n) = P(X_{n+1} = j \mid X_n = i) \ ; \ i, j, n = 0, 1, \ldots . \qquad (5.2)$$

From now on, we shall assume that the one-step transition probabilities do not depend on the time instant:

$$q_{i,j}(n) = q_{i,j} \ ; \ i, j, n = 0, 1, \ldots . \tag{5.3}$$

This restriction simplifies the treatment, without reducing the generality of the theory. Markov chains which satisfy (5.3) are said to be 'time-homogeneous'. That qualification will not be mentioned in future, but will be implied.

Thus, a Markov chain is characterized by its 'transition probability matrix', Q, containing the one-step transition probabilities:

$$Q = \begin{bmatrix} q_{0,0} & q_{0,1} & q_{0,2} & \cdots \\ q_{1,0} & q_{1,1} & q_{1,2} & \cdots \\ \vdots & \vdots & \vdots & \\ q_{i,0} & q_{i,1} & q_{i,2} & \cdots \\ \vdots & \vdots & \vdots & \end{bmatrix}, \tag{5.4}$$

where the indices range over the state space. Since the chain must be in some state at any observation instant, all row sums of the transition probability matrix are equal to 1:

$$\sum_{j=0}^{\infty} q_{i,j} = 1 \ ; \ i = 0, 1, \ldots . \tag{5.5}$$

Conversely, for any square matrix, Q (finite or infinite), whose elements are non-negative and satisfy (5.5), one can construct a Markov chain which has Q as its transition probability matrix.

A very convenient and visually appealing representation of a Markov chain is obtained by drawing a directed graph, with vertices corresponding to the states of the chain and arcs showing the non-zero one-step transition probabilities. That graph is called the 'state diagram' of the Markov chain.

Examples

1. A machine can be in one of two possible states: broken or operative (0 or 1, respectively). If it is broken at time n, it will be broken or operative at time $n+1$ with probabilities 0.3 and 0.7 respectively. If it is operative at time n, it will be broken or operative at time $n+1$ with probabilities 0.1 and 0.9 respectively ($n = 0, 1, \ldots$). This behaviour

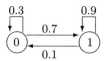

Fig. 5.2. An unreliable machine

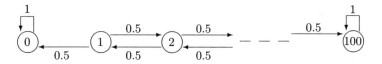

Fig. 5.3. The gambler's fortune

satisfies the Markov property. The transition probability matrix is

$$Q = \begin{bmatrix} 0.3 & 0.7 \\ 0.1 & 0.9 \end{bmatrix}.$$

The state diagram of this two-state Markov chain is shown in figure 5.2.

2. A gambler bets £1 at a time on the toss of a coin. She wins £1 if 'heads' comes up and loses her stake on 'tails'. She stops playing when her capital reaches either £0 or £100.

If $X_0 < 100$ is the gambler's initial capital and X_n is the amount she has after the nth move, $\{X_n : n = 0, 1, \ldots\}$ is a Markov chain. The transition probabilities have the following values ($i, j = 0, 1, \ldots, 100$):

$$Q = \begin{bmatrix} 1 & 0 & 0 & \ldots & 0 & 0 & 0 \\ 0.5 & 0 & 0.5 & \ldots & 0 & 0 & 0 \\ \vdots & \vdots & \vdots & & \vdots & \vdots & \vdots \\ 0 & 0 & 0 & \ldots & 0.5 & 0 & 0.5 \\ 0 & 0 & 0 & \ldots & 0 & 0 & 1 \end{bmatrix}.$$

The corresponding state diagram is shown in figure 5.3.

3. A discrete-time communication channel with a finite buffer of size N behaves as follows. During the nth slot, either a new message arrives (if there is room for it), with probability α, or one of the messages in the buffer (if any) is transmitted, with probability β, or there is no change. This system can be modelled by a Markov chain which is in state i when

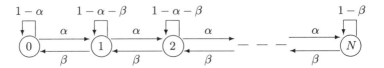

Fig. 5.4. A finite buffer

there are i messages in the buffer ($i = 0, 1, \ldots, N$). From state i, there are transitions to states $i-1$ (except when $i = 0$), i and $i+1$ (except when $i = N$), with probabilities β, $1 - \alpha - \beta$ and α, respectively. The state diagram is ilustrated in figure 5.4.

4. A population of single-cell organisms evolves according to the following simple rules. The lifespan of every cell is exactly one time unit. Then it either dies, with probability α, or divides into two new ones, with probability $1 - \alpha$, independently of all the other cells.

Denote by X_n the size of the population at time n (X_0 being the initial size). Then $\{X_n : n = 0, 1, \ldots\}$ is a Markov chain whose state space is the set of all non-negative even integers. Indeed, it is easy to see that although X_0 can be arbitrary, the values of $X_1, X_2, \ldots$ are always even. This is because the only possible one-step transitions from state i ($i = 1, 2, \ldots$) are to states $0, 2, 4, \ldots, 2i$. Assume, for simplicity, that X_0 is also even. Then the one-step transition probability from state $2i$ to state $2j$ ($j = 0, 1, \ldots, 2i$) is equal to the probability that j of the $2i$ cells divide and the other $2i - j$ cells die:

$$q_{2i,2j} = \binom{2i}{j} \alpha^{2i-j}(1-\alpha)^j \; ; \; j = 0, 1, \ldots, 2i \,.$$

* * *

Consider now the transitions made by the chain $X = \{X_n : n = 0, 1, \ldots\}$ in two steps. Since X is time-homogeneous, the corresponding probabilities are independent of the step index. Define

$$q_{i,j}^{(2)} = P(X_{n+2} = j \mid X_n = i) \; ; \; i, j, n = 0, 1, \ldots \,. \tag{5.6}$$

In order to move from state i to state j in two steps, the chain has to pass through some intermediate state, k, after the first step. Condition-

5.1 Markov chains

ing upon that intermediate state, we can write

$$q_{i,j}^{(2)} = \sum_{k=0}^{\infty} P(X_{n+2} = j \mid X_n = i, X_{n+1} = k) P(X_{n+1} = k \mid X_n = i)$$

$$= \sum_{k=0}^{\infty} P(X_{n+2} = j \mid X_{n+1} = k) q_{i,k} = \sum_{k=0}^{\infty} q_{i,k} q_{k,j}, \quad (5.7)$$

where the second equality uses the Markov property. Note that the right-hand side of (5.7) is the element (i, j) of Q^2. A simple extension of this argument leads to the following proposition:

Theorem 5.1 *Let* $q_{i,j}^{(s)}$ *be the s-step transition probability from state i to state j:*

$$q_{i,j}^{(s)} = P(X_{n+s} = j \mid X_n = i) \; ; \; s = 1, 2, \ldots . \quad (5.8)$$

The matrix of these probabilities, $Q^{(s)} = (q_{i,j}^{(s)})$, is given by

$$Q^{(s)} = Q^s .$$

This result may, in principle, be used to compute various quantities of interest associated with the performance of a Markov chain over a finite period of time. For instance, in example 2, we may wish to find (i) the probability that the gambler will lose all her money in not more than 50 moves, given that she started with £10, or (ii) the average sum of money in her possession after 50 moves, given that she started with £10. The answer to (i) is $q_{10,0}^{(50)}$, while (ii) is obtained from

$$E(X_{50} \mid X_0 = 10) = \sum_{j=1}^{100} j q_{10,j}^{(50)} .$$

For the particular transition probability matrix in the example, these tasks are not too difficult. However, similar computations for other models may be prohibitively expensive.

5.1.1 Steady state

Our main objective is the long-term analysis of Markov chains. Under certain conditions, the more steps the chain makes, the less it matters in what state it was when it started. In the limit, when the observation instant is infinitely far removed from the starting point, the probability of finding the chain in state j, p_j, is independent of the initial state:

$$\lim_{n\to\infty} P(X_n = j \mid X_0 = i) = \lim_{n\to\infty} q_{i,j}^{(n)} = p_j \ ; \ j = 0, 1, \ldots . \qquad (5.9)$$

When the limiting probabilities p_j exist, and add up to 1, they are referred to as the 'long-term distribution', or the 'steady-state distribution', or the 'equilibrium distribution' of the Markov chain. The problems of deciding whether a steady-state distribution exists, and of determining it if it does, are central to both the theory and the applications of Markov chains. We shall state some of the important results in this area, but in order to do that, a few definitions are necessary.

If, having once been in state i, there is a non-zero probability that the chain will eventually be in state j, then state j is said to be 'reachable' from state i. An equivalent and easier to use definition of the same concept can be given by means of the state diagram: if, starting in state i and following the transition arrows from state to state, one can arrive at state j, then j is reachable from i. This relation is not necessarily symmetric: j may be reachable from i without i being reachable from j.

In example 1, obviously both states are reachable from each other. In example 2, state 0 is reachable from every state except 100; similarly, state 100 is reachable from every state except 0. Having entered one of those two states, the chain remains there forever. Such states are called 'absorbing'. In example 3, every state is reachable from every state. In example 4, state 0 is reachable from every state and is absorbing; every even-numbered state is reachable from every non-zero state.

A set, C, of states is said to be 'closed' if no state outside C can be reached from a state in C. The set of all states is obviously closed. If that is the only closed set, i.e. if no proper subset of states is closed, then the Markov chain X is said to be 'irreducible'. It is readily seen that X is irreducible if, and only if, every state can be reached from every other state. The chains in examples 1 and 3 are irreducible; those in examples 2 and 4 are not.

If, having once been in state i, the chain will return to it eventually with probability 1, then state i is said to be 'recurrent'. Clearly, a recurrent state is entered by the Markov chain either not at all, or infinitely many times (since no visit to i can be the last one).

States which are not recurrent are called 'transient'. It can be shown that the total number of visits to a transient state is finite, with probability 1 (see exercise 2). Every transient state is eventually entered by

5.1 Markov chains

the Markov chain for the last time, and then no more. For instance, if C is a closed set of states and i is not in C, and if a state in C is reachable from i, then i is transient. With a non-zero probability, after visiting state i the chain will enter C and will not afterwards return to i. Such is the situation in example 2: the states $1, 2, \ldots, 99$ are transient. In the long run, the Markov chain will be trapped in one of the absorbing states, 0 or 100.

The long-term probability of any transient state is obviously 0.

If i is a recurrent state and j is a state reachable from i, then j is recurrent too. Indeed, suppose that j is transient and consider the path followed by the chain after a visit to j. A return to j is not certain. Hence, a visit to i is also not certain (if it were, then the chain would keep returning to i and would eventually find its way back to j). That means, however, that it is possible for the chain to move from i to j and then not return to i, which contradicts the fact that i is recurrent. Therefore, j is recurrent.

The above argument implies that, if a Markov chain is irreducible, then either all its states are transient or all its states are recurrent. We refer to these two cases by saying that the chain itself is transient, or recurrent, respectively.

Consider the path followed by an irreducible and recurrent Markov chain $X = \{X_n : n = 0, 1, \ldots\}$. Every state is visited by X infinitely many times, with probability 1, regardless of the initial state. Let $n_1^j, n_2^j, \ldots$ be the moments of consecutive visits to state j. Denote by m_j the expected interval between those visits:

$$m_j = E(n_{k+1}^j - n_k^j) \ . \tag{5.10}$$

If the chain makes a large number of steps, N, it will visit state j N/m_j times, on the average. In other words, X spends a fraction $1/m_j$ of its time in state j. Intuitively, that fraction should be equal to the long-term probability, p_j, of observing X in state j. This is indeed true, provided that the pattern of visits to the various states is not too regular.

A state j is said to be 'periodic', with period m ($m > 1$), if the consecutive returns to j occur only at multiples of m steps:

$$P(X_{n+s} = j \mid X_n = j) = 0 \text{ if } s \neq km \text{ for some } k \geq 1 \ . \tag{5.11}$$

If there is no integer $m > 1$ which satisfies (5.11), then j is said to be 'aperiodic'. If the Markov chain is irreducible, then either all its states are periodic, with the same period, or all of them are aperiodic. The chain itself is then said to be periodic, or aperiodic, respectively.

If an irreducible Markov chain has at least one state to which it can return in a single step, then it is aperiodic.

The first limiting result can now be stated:

Theorem 5.2 *If $X = \{X_n : n = 0, 1, \ldots\}$ is an irreducible, aperiodic and recurrent Markov chain, then the limiting probabilities p_j, defined in (5.9), exist and are given by*

$$p_j = \frac{1}{m_j} \; ; \; j = 0, 1, \ldots, \qquad (5.12)$$

where m_j is the average number of steps between consecutive visits to state j.

We shall omit the proof. Note that, although X keeps returning to a recurrent state with probability 1, it may do so at intervals whose average length is infinite. If $m_j = \infty$, then state j is said to be 'recurrent null'. According to the above theorem, $p_j = 0$ when j is recurrent null. The states whose average return times are finite, and whose long-term probabilities are therefore non-zero, are called 'recurrent non-null'.

It can be shown that if X is an irreducible and recurrent Markov chain, then either all its states are recurrent null, or all are recurrent non-null. Hence, either $p_j = 0$ for all j, or $p_j > 0$ for all j.

Relation (5.12) is quite instructive, but it does not tell us how to find the quantities m_j, and hence the probabilities p_j. Nor is it obvious, in general, whether the states of X are transient, recurrent null or recurrent non-null. For practical applications it is desirable to establish a connection between the one-step transition probabilities of a Markov chain and its steady-state distribution. This is provided by the following result, known as the 'steady-state theorem':

Theorem 5.3 *An irreducible and aperiodic Markov chain, X, with one-step transition probability matrix $Q = (q_{i,j})$, $i, j = 0, 1, \ldots$, is recurrent non-null if, and only if, the set of equations*

$$p_j = \sum_{i=0}^{\infty} p_i q_{i,j} \; ; \; j = 0, 1, \ldots, \qquad (5.13)$$

$$\sum_{j=0}^{\infty} p_j = 1, \qquad (5.14)$$

has a positive solution. That solution is then unique and is the steady-state distribution of X.

5.1 Markov chains

Equations (5.13) are referred to as the 'balance equations' of the Markov chain X, while (5.14) is the 'normalizing' equation. Of course, if the state space is finite, so are the sums in the right-hand sides of (5.13) and (5.14). Introducing the row vector $\mathbf{p} = (p_1, p_2, \ldots)$, equations (5.13) can be written in the form

$$\mathbf{p} = \mathbf{p}Q . \tag{5.15}$$

The balance equations have a simple intuitive interpretation which makes their necessity for the existence of a steady-state distribution almost obvious. Indeed, suppose that the limits (5.9) exist, and consider the chain at some moment in the steady state (i.e. after it has been running for a long time). At that moment, X is in state i with probability p_i and, if that is the case, the next state entered will be j with probability $q_{i,j}$. Hence, the right-hand side of (5.13) is equal to the unconditional probability that at the next observation instant the chain will be in state j. But the next instant is also in the steady state, which means that that probability must be equal to p_j.

To show the sufficiency of (5.13) and (5.14), let the initial state of the Markov chain be chosen at random, with the distribution p_j ($j = 0, 1, \ldots$). Then, since the balance equations are satisfied, the same distribution will be preserved at all subsequent observation instants, no matter how distant. That, however, implies that the Markov chain can be neither transient, nor recurrent null, for in both those cases all the long-run probabilities would be 0, contradicting (5.14).

The first limiting result, (5.12), tells us that a Markov chain cannot have more than one steady-state distribution. From this, and from the above argument, it follows that the solution to (5.13) and (5.14), if it exists, is unique.

Thus, the steady-state analysis of a system modelled by an irreducible and aperiodic Markov chain largely consists of solving the corresponding balance and normalizing equations. That may be an easy or a difficult task, depending on the size of the state space and on the structure of the one-step transition probability matrix.

When the state space is finite, the chain is always recurrent non-null (see exercise 4) and therefore (5.13) and (5.14) always have a solution. Numerically, that solution is usually obtained by discarding one of the balance equations and replacing it with the normalizing equation.

Note that the probability p_j appears in the right-hand side of (5.13), with a coefficient $q_{j,j}$. Bringing that term to the left-hand side, (5.13)

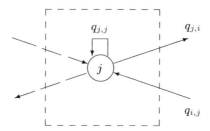

Fig. 5.5. Transitions into and out of a cut around state j

becomes

$$p_j(1 - q_{j,j}) = \sum_{i=0,\ i\neq j}^{\infty} p_i q_{i,j} \ . \tag{5.16}$$

Bearing in mind that $1-q_{j,j}$ is equal to the sum of all one-step transition probabilities *out of state j*, (5.16) can be written in the form

$$p_j \sum_{i=0,\ i\neq j}^{\infty} q_{j,i} = \sum_{i=0,\ i\neq j}^{\infty} p_i q_{i,j} \ . \tag{5.17}$$

Thus, a balance equation can be interpreted as saying that the average fraction of steps on which the chain makes a transition out of state j is equal to the average fraction of steps on which the chain makes a transition into state j. This is indeed what one would expect from a system in steady state.

Given a state diagram for the Markov chain, the balance equation (5.17) can be visualized by making an imaginary 'cut' around state j, as shown in figure 5.5. To each arc leading out of the cut there corresponds a term in the left-hand side of (5.17); similarly, to each arc leading into the cut there corresponds a term in the right-hand side.

Let us apply this approach to the Markov chains in examples 1 and 3 in this section. Both those chains are irreducible, because all their states can be reached from each other. Moreover, both chains are aperiodic, since they have states to which they can return after a single step. Therefore, a unique steady-state distribution exists in both cases.

1. The Markov chain modelling the unreliable machine has only two states. The corresponding steady-state probabilities can be determined from one of the balance equations (5.17) (make a cut round state 0 in the diagram in figure 5.2: there is one arc going out and one arc coming in), plus the normalizing equation:

$$0.7 p_0 = 0.1 p_1 \; ; \; p_0 + p_1 = 1 \; .$$

This yields $p_0 = 1/8$, $p_1 = 7/8$.

3. Making cuts round states $0, 1, \ldots, N$ in the diagram in figure 5.4, we obtain the following balance equations:

$$\begin{aligned} \alpha p_0 &= \beta p_1 \; , \\ (\alpha + \beta) p_j &= \alpha p_{j-1} + \beta p_{j+1} \; ; \; j = 1, 2, \ldots, N-1 \; , \\ \alpha p_{N-1} &= \beta p_N \; . \end{aligned}$$

The general solution of these equations is easily found, by consecutive elimination, to be of the form

$$p_j = \gamma^j p_0 \; ; \; j = 0, 1, \ldots, N \; ,$$

where $\gamma = \alpha/\beta$. The normalizing equation then yields

$$p_0 = \frac{1-\gamma}{(1-\gamma)^{N+1}} \; ,$$

thus determining all the steady-state probabilities. From these, one can compute various performance measures. For instance, the average number of messages in the buffer, L, is given by

$$L = \sum_{j=1}^{N} j \, p_j \; .$$

Exercises

1. Assuming that the gambler in example 2 starts playing with an initial capital of £2, what amounts could she have after three moves, and with what probabilities?

2. Let j be a transient state for some Markov chain. Denote by v_{jj} the probability that, after visiting j, the chain will eventually return to it. Show that, following an initial visit to state j, the total number of returns to it, K_j, is distributed geometrically:

$$P(K_j = k) = v_{jj}^{k-1}(1 - v_{jj}) \; ; \; k = 0, 1, \ldots \; .$$

Hence deduce that K_j is finite with probability 1.

3. Random walk. A particle meanders among the non-negative integer points on the x-axis. If, at time n, it is in position i ($i = 1, 2, \ldots$), then at time $n+1$ it can move to point $i-1$ with probability α, or to point $i+1$ with probability $1 - \alpha$. If it is at point 0, then it either remains there, with probability α, or moves to point 1 with probability $1 - \alpha$.

Draw the state diagram of the Markov chain $X = \{X_n : n = 0, 1, \ldots\}$, where X_n is the position of the particle at time n. Show that X is irreducible and aperiodic, and that the solution of the balance equations is of the form $p_j = \gamma^j p_0$, where $\gamma = (1-\alpha)/\alpha$. Hence demonstrate that X is recurrent non-null if, and only if, $\gamma < 1$, or $\alpha > 1/2$. In that case, the steady-state distribution of X is geometric:

$$p_j = \gamma^j (1 - \gamma) \; ; \; j = 0, 1, \ldots .$$

4. Consider an irreducible Markov chain $X = \{X_n : n = 0, 1, \ldots\}$, with a finite state space, $\{0, 1, \ldots, N\}$. Assume that X is transient, or recurrent null, and that the limiting probabilities of all states are therefore equal to 0. Show that this leads to a contradiction with the identity

$$\sum_{i=0}^{N} P(X_n = i) = 1 ,$$

which holds for every n, regardless of the initial state. Hence deduce that X is recurrent non-null.

5.2 A Markov chain embedded in the M/G/1 model

We shall examine an important and non-trivial application of a Markov chain with an infinite state space. The system of interest is the M/G/1 queue, introduced in chapter 3 (a number of average performance measures were derived there by elementary methods). Now we are able to determine the steady-state probabilities, p_j, that there are j jobs in the system, and also the probability density function, $g(x)$, of the response time.

The assumptions of the model are the same as in section 3.2. Jobs arrive in a Poisson stream with rate λ. The service times have some general distribution, with pdf $f(x)$ and mean b. The scheduling policy is FIFO.

One way of associating a stochastic process with this system is to define X_t as the number of jobs in the system at time t. Unfortunately,

5.2 A Markov chain embedded in the M/G/1 model

the process $\{X_t : t \geq 0\}$ does not have the Markov property. The next departure instant depends, in general, on the starting point of the current service, i.e. the future behaviour of X_t depends on its past history. It turns out, however, that a Markov chain can be 'embedded' in the process X_t. That insight was due to D.G. Kendall.

Let t_n be the instant when the nth job departs from the system ($n = 1, 2, \ldots$), and let X_n be the number of jobs in the system at time t_n^+, i.e. just after the nth departure. Clearly, if the value of X_n is known, then the value of X_{n+1} does not depend on anything that happened before t_n. Indeed, the arrivals following t_n are not influenced by past events because the arrival process is Poisson; the departures following t_n are not influenced by past events because there is no service in progress at t_n. All this implies that $X = \{X_n : n = 1, 2, \ldots\}$ is a Markov chain.

In the steady state, the distribution of the number of jobs left behind by a departure is the same as the distribution of the number of jobs seen by an arrival. This is because the fraction of departures that leave the system in state j is equal to the fraction of arrivals that see it in state j (see exercise 3 in section 3.2). On the other hand, the distribution of the number of jobs seen by an arrival is the same as the distribution of the number of jobs seen by a random observer, according to the PASTA property. Therefore, if we find the limiting probabilities of X, p_j ($j = 0, 1, \ldots$), we shall also have the steady-state distribution of the number of jobs seen in the system by a random observer.

Suppose that $X_n = i$, for $i \geq 1$, and consider the possible values of X_{n+1}. At time t_n, one of the i jobs in the system starts its service; that job departs at t_{n+1}. If j new jobs arrive during the service time, X_{n+1} will be equal to $i + j - 1$. If, on the other hand, $X_n = 0$, then t_n is the start of an idle period. Eventually, a job arrives into the system and starts service; that job departs at t_{n+1}. If j jobs arrive during the service, then X_{n+1} will be equal to j. These observations can be summarized by writing

$$X_{n+1} = \begin{cases} X_n + \zeta - 1 & \text{if } X_n > 0 \\ \zeta & \text{if } X_n = 0, \end{cases} \tag{5.18}$$

where ζ is the random variable representing the number of jobs that arrive during a service time. Denoting the latter's distribution by $r_k = P(\zeta = k)$, we can express the one-step transition probabilities of X as

$$q_{i,j} = P(X_{n+1} = j \mid X_n = i) = \begin{cases} r_{j-i+1} & \text{if } i > 0 \\ r_j & \text{if } i = 0, \end{cases} \tag{5.19}$$

170   5 Markov chains and processes

The Markov chain X is irreducible (because every state is reachable from every other state) and aperiodic (because every state is reachable from itself in one step). According to the steady-state theorem, X has a steady-state distribution if the following balance equations have a solution that can be normalized:

$$p_j = p_0 r_j + \sum_{i=1}^{j+1} p_i r_{j-i+1} \; ; \; j = 0, 1, \ldots . \tag{5.20}$$

The best way of dealing with this infinite set of equations is to transform it into a single equation by introducing the generating functions, $p(z)$ and $r(z)$, of the distributions p_j and r_k, respectively:

$$p(z) = \sum_{j=0}^{\infty} p_j z^j \; ; \; r(z) = \sum_{k=0}^{\infty} r_k z^k . \tag{5.21}$$

Multiplying the jth equation in (5.20) by z^j and summing over all j we get

$$p(z) = p_0 r(z) + \sum_{j=0}^{\infty} \sum_{i=1}^{j+1} p_i r_{j-i+1} z^j . \tag{5.22}$$

Replace, in the second term in the right-hand side of (5.22), z^j by $z^i z^{j-i+1} z^{-1}$, change the order or summation and separate the two sums. This yields

$$p(z) = p_0 r(z) + \frac{1}{z} r(z)[p(z) - p_0] . \tag{5.23}$$

The unknown generating function $p(z)$ can now be expressed in terms of $r(z)$ and p_0:

$$p(z) = \frac{p_0(1-z)r(z)}{r(z) - z} . \tag{5.24}$$

The distribution, and the generating function, of the number of Poisson arrivals during a random interval were derived in section 2.3.2. We repeat the relevant result here:

$$r(z) = f^*(\lambda - \lambda z) , \tag{5.25}$$

where $f^*(s)$ is the Laplace transform of the required service time probability density function. The latter is given as part of the model assumptions.

Substituting (5.25) into (5.24), we obtain

$$p(z) = \frac{p_0(1-z)f^*(\lambda - \lambda z)}{f^*(\lambda - \lambda z) - z} . \tag{5.26}$$

5.2 A Markov chain embedded in the M/G/1 model

It remains to establish whether the distribution provided by (5.26) can be normalized. For that, we must have $p(1) = 1$. Setting $z = 1$ in (5.26), remembering that $f^*(0) = 1$ and resolving the indeterminacy in the right-hand side by L'Hospital's rule, yields

$$p(1) = \frac{p_0}{1 + \lambda f^{*\prime}(0)} = \frac{p_0}{1 - \lambda b}, \qquad (5.27)$$

where b is the average service time. The conclusion is that the solution of the balance equations can be normalized if, an only if, the offered load, $\rho = \lambda b$, satisfies $\rho < 1$. Then $p_0 = 1 - \rho$. Of course, we already knew that from the arguments in chapter 3.

The average number of jobs in the system, L, is given by $p'(1)$. In working out the derivative in expression (5.26) at $z = 1$, one has to apply L'Hospital's rule twice. What emerges then is the Pollaczek–Khinchin formula of section 3.2. However, the present result is more powerful because it can be used to determine higher moments as well, by taking further derivatives at $z = 1$. The kth moment of the queue size depends on moments $1, 2, \ldots, k+1$ of the service time distribution.

Individual probabilities, p_j, can be computed by taking derivatives in (5.26) at $z = 0$.

To determine the distribution of the response time, we make the following simple observation. The jobs that are in the system at the moment of a job's departure are precisely the ones that arrived during the time that job spent in the system. Therefore, we can write a relation similar to (5.25), between the generating function of the number of jobs in the system and the Laplace transform, $g^*(s)$, of the density function of the response time. That relation is

$$p(z) = g^*(\lambda - \lambda z). \qquad (5.28)$$

Substituting (5.26) and making a change of variables, $s = \lambda - \lambda z$, we get

$$g^*(s) = \frac{(1 - \rho)sf^*(s)}{\lambda f^*(s) - \lambda + s}. \qquad (5.29)$$

By taking derivatives in (5.29) at $s = 0$, one can find the moments of the response time. The average was already determined in section 3.2.

In the special case when the service time distribution is exponential, with mean b, the Laplace transform $f^*(s)$ is equal to (see exercise 4 in section 2.2)

$$f^*(s) = \frac{1}{1 + bs}. \qquad (5.30)$$

Equation (5.26) now becomes

$$p(z) = \frac{1-\rho}{1-\rho z},$$

which is the generating function of the modified geometric distribution with parameter $(1-\rho)$. We thus rediscover the steady-state distribution of the M/M/1 queue (section 3.2.2):

$$p_j = (1-\rho)\rho^j \ ; \ j = 0, 1, \ldots .$$

Substituting (5.30) into (5.29) yields

$$g^*(s) = \frac{1-\rho}{1-\rho+bs},$$

which is the Laplace transform of the exponential distribution with mean $b/(1-\rho)$. Again, this confirms the result established in section 3.2.2.

5.3 Markov processes

We now turn our attention to the continuous time analogue of the Markov chain. This is a stochastic process, $X = \{X_t : t \geq 0\}$, whose time parameter takes arbitrary non-negative real values. The state space is still assumed to be discrete; it can be identified with the set, or a subset, of the non-negative integers. Continuous time processes are used to model systems where changes of state can occur at arbitrary moments, and where the intervals between those changes can be of arbitrary length.

The process X is a 'Markov process' if it has the Markov property. That is, the path followed by X after a given moment, t, depends only on the state at that moment, X_t, and not on the past history:

$$P(X_{t+y} = j \mid X_u : u \leq t) = P(X_{t+y} = j \mid X_t) \ ; \ j = 0, 1, \ldots . \quad (5.31)$$

The theory of Markov processes, especially as far as their long-term behaviour is concerned, has many parallels with that of Markov chains. In particular, much of the terminology of the previous section carries over almost unchanged.

A Markov process X is said to be 'time-homogeneous' if the right-hand side of (5.31) does not depend on the moment of observation, t, but is a function only of y, j and the value of X_t:

$$P(X_{t+y} = j \mid X_t = i) = q_{i,j}(y) \ ; \ i,j = 0,1,\ldots \ ; \ y \geq 0. \quad (5.32)$$

5.3 Markov processes

From now on, all Markov processes considered will be assumed to be time-homogeneous.

The functions (5.32) are called the 'transition probability functions' of the Markov process. They are analogous to the s-step transition probabilities of a Markov chain, defined in (5.8).

The evolution of a typical Markov process (excluding certain pathological cases) can be described as follows.

The process enters a state, say i ($i = 0, 1, \ldots$). It remains there for a random period of time, distributed exponentially with some parameter, a_i. At the end of that period, the process moves to a different state, j ($j = 0, 1, \ldots, j \neq i$), with some probability, $q_{i,j}$. It then remains in state j for a period of time distributed exponentially with parameter a_j, moves to state k, $k \neq j$), with probability $q_{j,k}$, etc.

This sort of behaviour is implied by the Markov property. Indeed, the latter implies that, if at any moment the process is observed in state i, the time that it will remain in that state is independent of the time already spent in it. But we know (see exercise 2 of section 2.2) that the only distribution which has this memoryless property is the exponential. Similarly, the next state to be entered depends only on the current state, and not on the time spent in it or on any previous states.

The significance of the above process structure is two-fold. First, the Markov property and the exponential distribution are very closely related. To assume that a system can be modelled by a Markov process is, in essence, equivalent to assuming that all intervals between events which change the system state are exponentially distributed.

The second important observation is that a Markov process is completely characterized by the parameters a_i and the transition probabilities $q_{i,j}$ ($i, j = 0, 1, \ldots : i \neq j$). In fact, we shall see that it is completely characterized by the products

$$a_{i,j} = a_i q_{i,j} \ ; \ i, j = 0, 1, \ldots \ ; \ i \neq j . \tag{5.33}$$

These quantities are called the 'instantaneous transition rates' of the Markov process. That name can be justified as follows.

Suppose that the process X is in state i at time t, and let h be a small time increment. What is the probability that the process will be in state j at time $t + h$? For that to happen, the period of residence in state i must terminate within the interval of length h and the process must move to state j (figure 5.6).

Remembering that the completion rate of an exponentially distributed

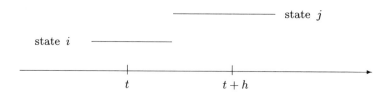

Fig. 5.6. Transition from state i to state j

random variable is equal to its parameter (section 2.2), we can write

$$P(X_{t+h} = j \mid X_t = i) = a_i h q_{i,j} + o(h) = a_{i,j} h + o(h) \ ; \ i \neq j, \quad (5.34)$$

where $o(h)$ is negligible compared to h (that term includes the probability of multiple transitions, e.g. from i to k and then from k to j, within the small interval h).

Thus, $a_{i,j}$ is the instantaneous rate at which the process moves from state i to state j, $i \neq j$. Alternatively, $a_{i,j}$ can be interpreted as the average number of transitions from state i to state j, per unit time spent in state i.

The instantaneous transition rates are the parameters of a Markov process model. They are either given as part of the model formulation, or are obtained from other basic assumptions. The object of any analysis is to derive performance measures in terms of these parameters.

Just as in the case of Markov chains, it is very convenient and useful to describe a Markov process by means of a state diagram. This is, once again, a directed graph where the nodes represent the process states. The arcs are now labelled with the corresponding instantaneous transition rates. Since a transition is, by definition, to a different state, there is never an arc from state i to state i.

Examples

1. Consider a Poisson arrival process with rate λ, and let K_t be the number of arrivals in the interval $(0, t)$. Then $K = \{K_t : t \geq 0\}$ is a Markov process, since the instants of future arrivals do not depend on those of past ones. If the process is in state i at time t, it will be in state $i + 1$ at time $t + h$ if one new arrival occurs between t and $t + h$; the probability of that event is $\lambda h + o(h)$ (see section 2.2). The probability of more than one arrival in an interval of length h is $o(h)$. Hence, the only

5.3 Markov processes

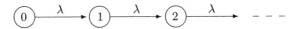

Fig. 5.7. State diagram of a Poisson process

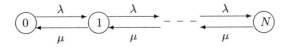

Fig. 5.8. State diagram of buffer process

non-zero instantaneous transition rates are $a_{i,i+1} = \lambda$ $(i = 0, 1, \ldots)$. The state diagram for the Poisson process is shown in figure 5.7.

2. (Continuous time version of the communication buffer from example 3 in section 5.1). Messages arrive into a buffer of size N in a Poisson stream with rate λ. Any new arrival which finds the buffer full is lost. Messages are removed from the buffer one at a time, independently of the arrivals, at intervals which are distributed exponentially with parameter μ.

Let X_t be the number of messages in the buffer at time t. The above assumptions ensure that $X = \{X_t : t \geq 0\}$ is a Markov process whose state space is the set $\{0, 1, \ldots, N\}$. If X is in state i at time t, the states to which it can move (with a non-negligible probability) within a short interval of length h are $i+1$ (if $i < N$ and a new message arrives between t and $t+h$) and $i-1$ (if $i > 0$ and a message is removed between t and $t+h$). The probabilities of those transitions are $\lambda h + o(h)$ and $\mu h + o(h)$, respectively. The probability of more than one arrival, or an arrival and a departure, is $o(h)$. Hence, the instantaneous transition rates of X are

$$a_{i,j} = \begin{cases} \lambda & \text{if } i < N \text{ and } j = i+1 \\ \mu & \text{if } i > 0 \text{ and } j = i-1 \\ 0 & \text{otherwise}. \end{cases}$$

The buffer process is illustrated in figure 5.8.

3. A very small telephone network connects four subscribers. Each of them attempts to make calls at intervals which are distributed exponentially with parameter λ, independently of the others. The recipient of a call is equally likely to be any of the other three subscribers, regardless of past history. If that recipient is not busy at the time, the call is

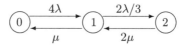

Fig. 5.9. Telephone network with four subscribers

successful and a conversation begins; otherwise the call is lost. The call durations are distributed exponentially with parameter μ.

As a state descriptor, we can use the number of conversations in progress at time t, X_t. Because of the exponential assumptions, $X = \{X_t : t > 0\}$ is a Markov process with state space $\{0, 1, 2\}$. Consider the possible transitions out of each of these three states.

In state 0, there are four unoccupied subscribers, each of which can make a call in a small interval of length h with probability $\lambda h + o(h)$. If anyone does, the call will be successful, since the recipient is known to be free. Hence, the probability that the process will move to state 1 is $4\lambda h + o(h)$. In state 1, there is one conversation in progress and two free subscribers. An attempt will be made to make a call in $(t, t + h)$ with probability $2\lambda h + o(h)$. However, such an attempt will be successful only if it is addressed to the other free subscriber, which will happen with probability $1/3$. Hence, the process will move to state 2 with probability $(2/3)\lambda h + o(h)$. It can also move to state 0, if the conversation which is in progress terminates. This will happen with probability $\mu h + o(h)$. In state 2, there are two conversations in progress and no free subscribers. Since either conversation can terminate, with probability $\mu h + o(h)$, the probability that the process will move to state 1 is $2\mu h + o(h)$. All other transitions occur with probability $o(h)$.

Thus, the instantaneous transition rates for this process are (see figure 5.9):

$$a_{0,1} = 4\lambda \; ; \; a_{1,2} = \frac{2}{3}\lambda \; ; \; a_{1,0} = \mu \; ; \; a_{2,1} = 2\mu .$$

* * *

Let us return now to the definitions (5.33) of the instantaneous transition rates. Since the sum of the transition probabilities $q_{i,j}$ over all j ($j \neq i$) must be 1, we can express the parameter a_i in terms of $a_{i,j}$:

$$a_i = \sum_{j=0,\ j\neq i}^{\infty} a_{i,j} \; ; \; i = 0, 1, \ldots . \tag{5.35}$$

5.3 Markov processes

In view of (5.33) and (5.35), the transition probabilities, $q_{i,j}$, can also be expressed in terms of $a_{i,j}$:

$$q_{i,j} = \frac{a_{i,j}}{a_i} = a_{i,j} \left[\sum_{j=0,\, j\neq i}^{\infty} a_{i,j} \right]^{-1}. \tag{5.36}$$

It is perhaps worth summarizing the significance of the parameters a_i and $a_{i,j}$ of a Markov process X:

1. When X enters state i, it remains there for an exponentially distributed interval with mean length $1/a_i$.
2. After leaving state i, X enters state j with probability $a_{i,j}/a_i$.
3. The instantaneous rate at which X leaves state i is a_i; this is also the average number of transitions out of state i per unit time spent in state i.
4. The instantaneous rate at which X moves from state i to state j is $a_{i,j}$; this is also the average number of transitions from state i to state j per unit time spent in state i.

Another way of describing the evolution of a Markov process is to imagine independent and exponentially distributed intervals with parameters $a_{i,j}$ ($j = 0, 1, \ldots$), all starting (or in progress) when X enters state i. The process remains in state i until the shortest of those intervals terminates; if the shortest interval is the one with parameter $a_{i,j}$, then the next state to be entered is state j. According to the results obtained in section 2.2.1, this interpretation is equivalent to the one summarized in points 1 and 2 above. Indeed, the shortest interval is distributed exponentially with parameter a_i and the probability that it is the one with parameter $a_{i,j}$ is equal to $a_{i,j}/a_i$.

Define the square matrix, A (with rows and columns indexed by states), whose off-diagonal elements are $a_{i,j}$, and whose diagonal ones are equal to $-a_i$:

$$A = \begin{bmatrix} -a_0 & a_{0,1} & a_{0,2} & \cdots \\ a_{1,0} & -a_1 & a_{1,2} & \cdots \\ \vdots & & & \\ a_{i,0} & \cdots & -a_i & \cdots \\ \vdots & & & \ddots \end{bmatrix}. \tag{5.37}$$

Equation (5.35) implies that every row sum of A is 0.

The matrix A is known as the 'generator' of the Markov process X,

178   5 Markov chains and processes

for reasons which will be explained shortly. Everything we may wish to know about X is determined, at least in principle, by A.

5.3.1 Transient behaviour

The behaviour of a Markov process over finite intervals of time is described by the transition probability functions, $q_{i,j}(t)$, specifying the probability that the process is in state j at time t, given that it started in state i at time 0:

$$q_{i,j}(t) = P(X_t = j \mid X_0 = i) \; ; \; i,j = 0,1,\ldots . \qquad (5.38)$$

Because of the Markov property, these are exactly the same functions as the ones in (5.32). They are also referred to as the 'transient distribution' of the Markov process.

We exclude from consideration processes that can leave a state immediately upon entering it (i.e. remain in a state for an interval of length 0). Then the transition probability functions satisfy

$$q_{i,j}(0) = \begin{cases} 1 & \text{if } i = j \\ 0 & \text{if } i \neq j . \end{cases} \qquad (5.39)$$

Moreover, the derivatives of $q_{i,j}(t)$ at $t = 0$ are given by the instantaneous transition rates:

$$q'_{i,i}(0) = -\lim_{h \to 0} \frac{1 - q_{i,i}(h)}{h} = -\lim_{h \to 0} \frac{a_i h + o(h)}{h} = -a_i , \qquad (5.40)$$

$$q'_{i,j}(0) = \lim_{h \to 0} \frac{q_{i,j}(h)}{h} = \lim_{h \to 0} \frac{a_{i,j} h + o(h)}{h} = a_{i,j} \; ; \; i \neq j . \qquad (5.41)$$

Introducing the matrix of transition probability functions, $Q(t) = [q_{i,j}(t)]$, we can express the above initial conditions as follows.

$$Q(0) = I \; ; \; Q'(0) = A , \qquad (5.42)$$

where I and A are the identity and generator matrices, respectively.

Now consider the probability that the process is in state j at time $t+x$, for some $x > 0$, given that it started in state i at time 0. On the way from i to j, the process must pass through some intermediate state, k, at time t. Moreover, given that intermediate state, the path followed between t and $t + x$ is independent of the one between 0 and t. Hence, the transition probability functions satisfy the so-called 'Chapman–Kolmogorov equations':

$$q_{i,j}(t+x) = \sum_{k=0}^{\infty} q_{i,k}(t) q_{k,j}(x) \ . \tag{5.43}$$

In matrix form, (5.43) can be written as

$$Q(t+x) = Q(t)Q(x) \ . \tag{5.44}$$

Taking the derivative in (5.44) with respect to x, at $x = 0$, yields a set of differential equations for the transition probability functions:

$$Q'(t) = Q(t)A \ . \tag{5.45}$$

These are known as the 'Kolmogorov forward differential equations'. Similarly, writing (5.44) as $Q(x+t) = Q(x)Q(t)$ and differentiating with respect to x at $x = 0$ produces another set of equations,

$$Q'(t) = AQ(t) \ , \tag{5.46}$$

known as the 'Kolmogorov backward differential equations'.

Either (5.45) or (5.46) can be solved for the transition probability functions, subject to the initial conditions $Q(0) = I$. If, instead of being a matrix, $Q(t)$ were an ordinary scalar function and A were a scalar constant, then the solution would be

$$Q(t) = e^{At} \ . \tag{5.47}$$

It turns out that this is a correct matrix solution to the Kolmogorov differential equations, provided that (5.47) is interpreted as

$$Q(t) = \sum_{n=0}^{\infty} \frac{t^n}{n!} A^n \ ; \ t \geq 0 \ . \tag{5.48}$$

Thus the matrix A, which describes how the process behaves in the neighbourhood of 0, determines its distribution at all times. That explains the term 'generator matrix'.

Expression (5.48) is not normally used for the purpose of computing the transient distribution of a Markov process. When the state space is finite, of size N, spectral methods are quite efficient: the transition probability functions satisfying Kolmogorov's equations are of the form

$$q_{i,j}(t) = \sum_{k=1}^{N} c_{i,j,k} e^{-z_k t},$$

where z_k are the eigenvalues of the generator matrix, and the coefficients $c_{i,j,k}$ are determined by the left and right eigenvectors. There are various numerical packages for solving sets of differential equations. However, when the state space is very large, or infinite, the usual approach is to seek approximate solutions.

5.3.2 Steady state

Our primary concern is with the long-term behaviour of a Markov process $X = \{X_t : t \geq 0\}$. The generator matrix, A, is assumed known. After the process has been running for a long time, the probability of observing it in state j should be independent of the initial state and of the observation moment. Denote that limiting probability by p_j:

$$p_j = \lim_{t \to \infty} P(X_t = j \mid X_0 = i) = \lim_{t \to \infty} q_{i,j}(t) \; ; \; j = 0, 1, \ldots . \quad (5.49)$$

The analysis involved in establishing the existence of the limits (5.49), and in determining them, is very similar to the corresponding one for the case of a Markov chain. This is not surprising, when one considers the close relationship between a Markov process and a Markov chain. Indeed, if we ignore the times at which the process X moves from state to state, and simply number those transitions, then the resulting sequence of states, $\{X_n : n = 0, 1, \ldots\}$, is a Markov chain. That chain is said to be 'embedded' in the process X. The embedded Markov chain has the same state space and the same initial state as the process X. Its one-step transition probabilities, $q_{i,j}$, are given by (5.36).

A Markov process is said to be irreducible, transient, recurrent null or recurrent non-null if its embedded Markov chain has those attributes. The question of periodicity does not arise in the case of the process because of the continuous time parameter.

There are two possibilities for the limiting probabilities of an irreducible Markov process.

(a) If the process is transient or recurrent null, then the limits p_j are equal to 0 for all j. When that is the case, we say that a steady-state distribution does not exist.

5.3 Markov processes

(b) If the process is recurrent non-null, then the limits p_j are positive for all j and sum up to 1. In that case, they constitute the steady-state distribution of the process.

To decide whether (a) or (b) holds, and to determine the steady-state distribution when it exists, we use the 'steady-state theorem for Markov processes':

Theorem 5.4 *An irreducible Markov process X, with generator matrix A, is recurrent non-null if, and only if, the set of equations*

$$p_j a_j = \sum_{i=0,\ i\neq j}^{\infty} p_i a_{i,j}\ ;\ j = 0, 1, \ldots, \qquad (5.50)$$

$$\sum_{j=0}^{\infty} p_j = 1. \qquad (5.51)$$

has a positive solution. That solution is then unique and is the steady-state distribution of X.

This result is very similar to the steady-state theorem for Markov chains. As in that case, (5.50) and (5.51) are called the 'balance equations' and 'normalizing equation', respectively. Introducing the row vector $\mathbf{p} = (p_0, p_1, \ldots)$, the balance equations can be written in the form

$$\mathbf{p}A = 0. \qquad (5.52)$$

Note that if both sides of (5.50) are divided by a_j, the result is precisely the balance equation of the embedded Markov chain. Thus, the existence of a steady-state distribution for the Markov process is equivalent to that for the embedded Markov chain.

Since a_j is the total instantaneous transition rate out of state j, the left-hand side of (5.50) can be interpreted as the average number of transitions out of state j per unit time. Similarly, the term $p_i a_{i,j}$ in the right-hand side is the average number of transitions from state i to state j per unit time. Clearly, the total average numbers of transitions out of, and into, state j per unit time must be equal in the steady state.

A visual representation of the balance equations in terms of the state diagram of the Markov process is obtained by substituting (5.35) into (5.50):

$$p_j \sum_{i=0,\ i\neq j}^{\infty} a_{j,i} = \sum_{i=0,\ i\neq j}^{\infty} p_i a_{i,j}\ ;\ j = 0, 1, \ldots. \qquad (5.53)$$

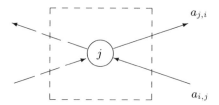

Fig. 5.10. Balance equation for a Markov process

Make a cut in the diagram around state j, as in figure 5.10. To each arc leading out of the cut there corresponds a term in the left-hand side of (5.53), while for each arc leading into the cut there is a term in the right-hand side.

It is also possible to make cuts around groups of several states, and equate the total average number of transitions out of the cut per unit time with the total average number of transitions into the cut. The resulting equations are not independent of the equations (5.53), but they may well be simpler and more convenient to work with.

As in the case of Markov chains, an irreducible Markov process with a finite state space is always recurrent non-null.

Example

4. Let us find the steady-state distribution of the Markov process modelling the four-subscriber telephone network (example 3 in the previous section). That process is irreducible and finite, so its steady-state distribution must exist.

Making cuts around states 0, 1 and 2 (see figure 5.9) yields the following balance equations:

$$\begin{aligned} 4\lambda p_0 &= \mu p_1 , \\ \left(\mu + \frac{2\lambda}{3}\right) p_1 &= 4\lambda p_0 + 2\mu p_2 , \\ \frac{2\lambda}{3} p_1 &= 2\mu p_2 . \end{aligned}$$

Solving, say, the first and the third of these, together with the normalizing equation, $p_0 + p_1 + p_2 = 1$, we get

$$p_0 = \left(1 + \frac{4\lambda}{\mu} + \frac{4\lambda^2}{3\mu^2}\right)^{-1} \;;\; p_1 = \frac{4\lambda}{\mu}p_0 \;;\; p_2 = \frac{4\lambda^2}{3\mu^2}p_0.$$

T   ‾age number of conversations in progress is given by $p_1 + 2p_2$.

 .e balance equations for the Poisson process K of example 1
 $($ $)$. Can their solution be normalized? Is K transient, recurrent
null   recurrent non-null?

2. In order to complete normally, a job has to go through M consecutive execution phases (e.g. compilation, linking, loading, etc.). The duration of phase i is exponentially distributed with parameter μ_i ($i = 1, 2, \ldots, M$). After completing phase i, the job starts phase $i+1$ with probability α_i (for $i < M$), or is aborted with probability $1 - \alpha_i$. An aborted job, or one which completes phase M, is replaced immediately with a new job which starts phase 1.

Let X_t be the index of the phase which is being executed at time t. Show that $X = \{X_t : t \geq 0\}$ is an irreducible Markov process with state space $\{1, 2, \ldots, M\}$. Find the steady-state distribution of X. Hence obtain the average number of jobs that complete normally, and the average number that are aborted, per unit time.

3. Parts arrive for processing at a manufacturing stage in a Poisson process with rate λ. There are two machines, each of which can process one part at a time, independently of the other. However, one of the machines works only when the total number of parts present is odd. If a change from odd to even occurs while a service is in progress at that machine (e.g. a new part arrives, or one departs from the other machine), the service is interrupted, to be resumed from the point of interruption when the number becomes odd again. Service times at both machines are distributed exponentially with parameter μ. The queue is unbounded.

Let X_t be the number of parts present at time t. Show that this is an irreducible Markov process with state space $\{0, 1, \ldots\}$. Write the balance equations and establish the condition for existence of steady state. Explain why the intuitive answer, $\rho < 1.5$, where $\rho = \lambda/\mu$, is not correct.

5.4 First passages and absorptions

There are important measures of system performance and reliability that concern the passage of a stochastic process, X, from a given initial state to a particular state or group of states. Let S be a proper and non-empty subset of states, and i be a state outside S. The interval between a moment when X enters state i, and the first subsequent moment when it enters one of the states in S, is called the 'first passage time' from i to S. If the states in S are absorbing, that interval is also called the 'time to absorption' from i to S. If X is a chain, the first passage time is an integer—the number of steps taken from entering i to entering S.

Below are some examples of first passage times that may be of interest.

1. The number of bets placed on a roulette wheel until a given initial capital is exhausted.
2. The time it takes for a traffic jam of a given size to dissipate.
3. The number of generations needed for a given population to double its size (alternatively, to become extinct).
4. The operative period of a machine, defined as the interval until more than a given number of its components have failed.
5. The number of tests needed to disclose a given number of faults in a software product.

Consider a Markov chain, X, with one-step transition probabilities $q_{i,j}$. Denote the average first passage time from state i to a set of states, S, by $m_{i,S}$ ($i \notin S$). These averages satisfy a set of linear equations, obtained as follows.

One step after entering state i, the chain will be in some state, j, with probability $q_{i,j}$. If j is not in S, then the remaining average number of steps until entering S is $m_{j,S}$. Hence we can write

$$m_{i,S} = 1 + \sum_{j \notin S} q_{i,j} m_{j,S} \; ; \; i \notin S . \tag{5.54}$$

If the state space is finite, and if there are no absorbing states outside S but reachable from i, these equations can be solved for the average first passage times. Otherwise, they may or may not have a finite solution.

Example

1. Let X be the Markov chain illustrated in figure 5.3 (gambler's fortune example in section 5.1). The set of absorbing states is $S = \{0, 100\}$. To

5.4 First passages and absorptions

find the average number of steps from an initial capital, i, to absorption we need to solve the equations

$$m_{1,S} = 1 + 0.5 m_{2,S},$$
$$m_{i,S} = 1 + 0.5 m_{i-1,S} + 0.5 m_{i+1,S} \; ; \; i = 2, 3, \ldots, 98,$$
$$m_{99,S} = 1 + 0.5 m_{98,S}.$$

The solution is finite. However, if S contains only one of the absorbing states, say $S = \{0\}$, then $m_{i,S} = \infty$ for all $i = 1, 2, \ldots, 99$. This is because, with a non-zero probability, the chain may reach state 100 first and remain there forever. Also, if the upper bound on the gambler's capital is removed, making the state space infinite, then $m_{i,0} = \infty$ for all $i = 1, 2, \ldots$.

* * *

In models with more than one absorbing state, we may be interested in the probability, $r_{i,k}$, that after visiting state i, the chain will be absorbed in state k. These probabilities satisfy the equations

$$r_{i,k} = q_{i,k} + \sum_{j \neq k} q_{i,j} r_{j,k} \; ; \; i \neq k. \tag{5.55}$$

In the above example, the probabilities $r_{i,0}$, of being ruined when starting with capital i, are found from

$$r_{1,0} = 0.5 + 0.5 r_{2,0},$$
$$r_{i,0} = 0.5 r_{i-1,0} + 0.5 r_{i+1,0} \; ; \; i = 2, 3, \ldots, 98,$$
$$r_{99,0} = 0.5 r_{98,0}.$$

Suppose now that X is a Markov process, with generator matrix A. Again, the average first passage time from state i to a set of states, S, is denoted by by $m_{i,S}$ ($i \notin S$). The equations for these averages are similar to (5.54).

After entering state i, the process remains there for an average interval $1/a_i$; then it moves to state j with probability $a_{i,j}/a_i$. Hence,

$$m_{i,S} = \frac{1}{a_i} + \sum_{j \notin S} \frac{a_{i,j}}{a_i} m_{j,S} \; ; \; i \notin S. \tag{5.56}$$

Example

2. A system consists of three identical machines, each of which goes through alternating periods of being operative and inoperative, indepen-

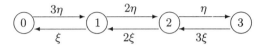

Fig. 5.11. Three unreliable machines

dently of the others. Those periods are distributed exponentially, with parameters ξ and η, respectively. The system as a whole is operative if at least two machines are operative.

The number of operative machines at time t, X_t, is a Markov process with state space $\{0, 1, 2, 3\}$. When X_t is in state i, there are i machines which may break down and $3-i$ machines which may become operative. Therefore, X_t moves to state $i-1$ with instantaneous rate $i\xi$ and to state $i+1$ with instantaneous rate $(3-i)\eta$. The state diagram is illustrated in figure 5.11.

Let $S = \{0, 1\}$ be the set of states where the system is not operative. Of interest are the average first passage times from the operative system states, 2 and 3, to S. The corresponding equations are

$$m_{3,S} = \frac{1}{3\xi} + m_{2,S},$$

$$m_{2,S} = \frac{1}{2\xi + \eta} + \frac{\eta}{2\xi + \eta} m_{3,S}.$$

These yield the following averages:

$$m_{2,S} = \frac{1}{2\xi} + \frac{\eta}{6\xi^2} \ ; \ m_{3,S} = \frac{1}{2\xi} + \frac{1}{3\xi} + \frac{\eta}{6\xi^2}.$$

5.5 Birth-and-Death queueing models

A large class of queueing systems are modelled by Markov processes of the so-called 'Birth-and-Death' type. The defining characteristic of these processes is that the only non-zero instantaneous transition rates out of state i $(i = 0, 1 \ldots)$ are to the neighbouring states (if any), $i+1$ and $i-1$. In the context of population models, these transitions can be thought of as a birth and a death, respectively, which explains the name. In the queueing framework, the process, X_t, represents the number of jobs present in some system at time t. Transitions from one state to another occur when jobs arrive and when they depart. Then the above condition amounts to a requirement that jobs arrive and depart singly (rather than in batches).

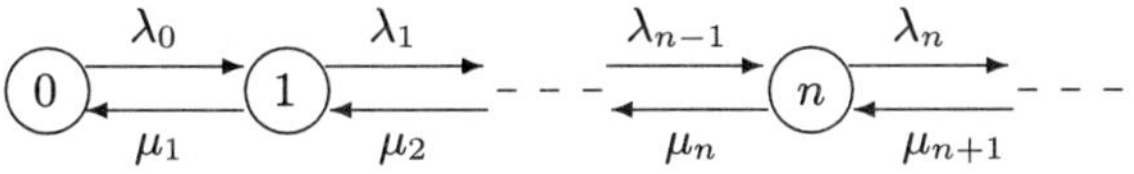

5.5 Birth-and-Death queueing models

Fig. 5.12. State diagram for the Birth-and-Death process

Denote the arrival rate $(a_{i,i+1})$ and departure rate $(a_{i,i-1})$, when the process is in state i, by λ_i and μ_i respectively. The state diagram for this process is illustrated in figure 5.12.

Denote the limiting probability of state j by p_j $(j = 0, 1, \ldots)$. These probabilities satisfy the following balance equations (the jth equation balances the average number of transitions out of, and into, a cut around state j):

$$(\lambda_j + \mu_j)p_j = \lambda_{j-1}p_{j-1} + \mu_{j+1}p_{j+1} \; ; \; j = 0, 1, \ldots,$$

where $\mu_0 = 0$ and $\lambda_{-1} = p_{-1} = 0$ by definition. An equivalent, but more convenient, set of equations is obtained by balancing the average number of transitions out of, and into, a cut enclosing the set of states $\{0, 1, \ldots, j-1\}$:

$$\lambda_{j-1}p_{j-1} = \mu_j p_j \; ; \; j = 1, 2, \ldots, n \,. \tag{5.57}$$

By successive elimination, all limiting probabilities can be expressed in terms of p_0:

$$p_j = \frac{\lambda_0 \lambda_1 \ldots \lambda_{j-1}}{\mu_1 \mu_2 \ldots \mu_j} p_0 \; ; \; j = 0, 1, \ldots, \tag{5.58}$$

where an empty product is 1 by definition.

Steady state exists for the Birth-and-Death process if, and only if, the solution (5.58) can be normalized, i.e. if it can be made to satisfy

$$\sum_{j=0}^{\infty} p_j = 1 \,.$$

Therefore, the condition for stability is that the value of p_0 obtained from

$$p_0 = \left[\sum_{j=0}^{\infty} \frac{\lambda_0 \lambda_1 \ldots \lambda_{j-1}}{\mu_1 \mu_2 \ldots \mu_j} \right]^{-1}, \tag{5.59}$$

is positive. In other words, the series in the right-hand side of (5.59) must

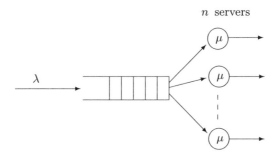

Fig. 5.13. The M/M/n queue

converge. Then the steady-state distribution of the process is given by (5.58) and (5.59).

A classic example of a Birth-and-Death process is the M/M/1 queue. Its birth and death rates are $\lambda_j = \lambda$ for all $j = 0, 1, \ldots$, and $\mu_j = \mu$ for $j = 1, 2, \ldots$ (the job departure rate is the rate at which a service completes, i.e. the parameter of the service time distribution). However, this model has been analysed twice already (in sections 3.2.2 and 5.2); there is not much point in deriving the same results by yet another method. We shall consider instead some models with multiple parallel servers and/or multiple sources of jobs.

5.5.1 The M/M/n model

In the M/M/n model, jobs arrive according to a Poisson process with rate λ. Service is provided by n identical parallel servers, each of which can serve one job at a time, independently of the others. All service times are distributed exponentially with parameter μ (the average service time is $b = 1/\mu$). The queue is unbounded and the scheduling policy is FIFO (figure 5.13).

It is easily seen that $X = \{X_t : t \geq 0\}$, where X_t is the number of jobs in the system at time t, is an irreducible Markov process whose states are the non-negative integers $\{0, 1, \ldots\}$. The instantaneous transition rate from state j to state $j + 1$ is equal to the arrival rate, $\lambda_j = \lambda$, for all $j = 0, 1, \ldots$. To determine the departure rates, consider the two cases $j < n$ and $j \geq n$.

For $j < n$, all j jobs in the system are being served and there are $n - j$ idle servers. Since each service completes in a small interval of length h

5.5 Birth-and-Death queueing models

with probability $\mu h + o(h)$, the probability that there is a departure in that interval is equal to $j\mu h + o(h)$. Hence, the instantaneous departure rate when the system is in state j is equal to $\mu_j = j\mu$, for $j = 1, 2, \ldots, n-1$.

When $j \geq n$, there are n busy servers, i.e. n services in progress. The instantaneous departure rate is now independent of j and is equal to $\mu_j = n\mu$, for all $i = n, n+1, \ldots$.

The general solution of the balance equations, (5.58), can now be written in the form

$$p_j = \begin{cases} (\rho^j/j!)p_0 & \text{for } j = 0, 1, \ldots, n-1 \\ (\rho^j/(n!n^{j-n}))p_0 & \text{for } j = n, n+1, \ldots, \end{cases} \quad (5.60)$$

where $\rho = \lambda/\mu$ is the offered load.

The service capacity of this system is n units of service per unit time (since there are n servers). Therefore, we can guess that the condition for stability is $\rho < n$. This is indeed the case. The expression for p_0, (5.59), is

$$p_0 = \left[\sum_{j=0}^{n-1} \frac{\rho^j}{j!} + \frac{\rho^n}{n!} \sum_{i=0}^{\infty} \left(\frac{\rho}{n}\right)^i \right]^{-1}.$$

This yields a positive value if, and only if, $\rho < n$. When that condition holds, we get

$$p_0 = \left[\sum_{j=0}^{n-1} \frac{\rho^j}{j!} + \frac{\rho^n}{(n-1)!(n-\rho)} \right]^{-1}. \quad (5.61)$$

The average number of jobs in the system, L, is equal to

$$L = \sum_{j=1}^{\infty} j p_j = \left[\sum_{j=1}^{n-1} \frac{\rho^j}{(j-1)!} + \frac{\rho^n(n^2 - n\rho + \rho)}{(n-1)!(n-\rho)^2} \right] p_0. \quad (5.62)$$

When $n = 1$, these expressions coincide with the M/M/1 results from chapter 3.

Example

1. Let us compare the steady-state performance of three systems: (a) an M/M/1 queue with parameters λ and 2μ, (b) an M/M/2 queue with parameters λ and μ and (c) two independent M/M/1 queues with parameters $\lambda/2$ and μ each. Note that since the total arrival rate, λ, and

the total service capacity, 2μ, are the same in all three cases, so are the conditions for stability: $\rho = \lambda/\mu < 2$. We shall use the average number of jobs, L, as the criterion for comparison.

For the M/M/1 system with offered load $\rho/2$ we have

$$L_{(a)} = \frac{\rho}{2-\rho}.$$

The M/M/2 expressions (5.61) and (5.62) yield

$$p_0 = \frac{2-\rho}{2+\rho},$$

and

$$L_{(b)} = \frac{4\rho}{(2-\rho)(2+\rho)}.$$

The offered load at each of the independent M/M/1 queues in model (c) is $\rho/2$. Therefore, the total average number of jobs in that system is

$$L_{(c)} = \frac{2\rho}{2-\rho}.$$

Bearing in mind that $1 < 4/(2+\rho) < 2$, we conclude that

$$L_{(a)} < L_{(b)} < L_{(c)}.$$

In general, for a given total arrival rate and total service capacity, an M/M/1 system is more efficient than an M/M/n one, which is itself more efficient than n independent M/M/1 queues. The reason is that the processing power of an n-server system is utilized fully only when all servers are busy, which requires at least n jobs in the M/M/n system, and at least n non-empty queues in system (c).

* * *

Let q_j be the conditional steady-state probability that there are j jobs *in the queue*, given that all servers are busy ($j = 0, 1, \ldots$). These probabilities are given by

$$q_j = \frac{p_{n+j}}{\sum_{i=0}^{\infty} p_{n+i}}. \qquad (5.63)$$

On the other hand, relations (5.60) imply that

$$p_{n+j} = \left(\frac{\rho}{n}\right)^j p_n \; ; \; j = 0, 1, \ldots . \qquad (5.64)$$

5.5 Birth-and-Death queueing models

Substitution of (5.64) into (5.63) yields, after cancellation of p_n and summation of the geometric series,

$$q_j = \left(\frac{\rho}{n}\right)^j \left(1 - \frac{\rho}{n}\right) \; ; \; j = 0, 1, \ldots . \tag{5.65}$$

Thus, the conditional distribution of the M/M/n queue size given that all servers are busy is the same as the unconditional distribution of the number of jobs in an M/M/1 system with offered load ρ/n.

Consider now the response time, Y, of a job in the M/M/n system. It is convenient to regard this as the sum of two independent random variables, $Y = Y_0 + Y_1$, where Y_0 is the waiting time (time spent in the queue) and Y_1 is the service time. The latter is distributed exponentially, with mean $1/\mu$. The problem is thus to find the distribution function, $H(x)$, of Y_0.

Denote by q the probability that a job has to wait, i.e. that it finds at least n jobs in the system on arrival. From (5.60) and (5.61) it follows that

$$q = P(Y_0 > 0) = \sum_{j=n}^{\infty} p_j = \left[1 + (n-1)!(n-\rho) \sum_{j=0}^{n-1} \frac{\rho^{j-n}}{j!} \right]^{-1} . \tag{5.66}$$

This expession is known as the 'Erlang's delay formula'.

Conditioning the waiting time distribution upon not having, or having to wait, we can write, for $x \geq 0$,

$$H(x) = P(Y_0 \leq x) = 1 - q + qP(Y_0 \leq x \,|\, Y_0 > 0) . \tag{5.67}$$

Now, while all servers are busy, jobs depart from the system at intervals which are exponentially distributed with parameter $n\mu$ (shortest of n exponential random variables). This fact, together with (5.65), implies that as far as a waiting job is concerned, the queue behaves as if it is served by a single exponential server with parameter $n\mu$. Hence, the conditional waiting time of a job that has to wait in the M/M/n system has the same distribution as the unconditional response time of a job in an M/M/1 system with parameters λ and $n\mu$. Substituting the expression from section 3.2.2 into (5.67) we get

$$H(x) = 1 - q + q[1 - e^{-(n\mu - \lambda)x}] \; ; \; x \geq 0 . \tag{5.68}$$

Note that this distribution function has a jump of size $1-q$ at the origin.

The average waiting time, w, is now easily derived from (5.68):

$$w = \frac{q}{n\mu - \lambda} . \tag{5.69}$$

The distribution function of the response time is the convolution of (5.68) with the exponential distribution function with parameter μ. The average response time, W, can be obtained either from $W = w + 1/\mu$ or from Little's theorem, $L = \lambda W$.

5.5.2 The M/M/n/n and M/M/∞ systems

One of the first 'queueing' systems ever analysied was the famous 'Erlang model' of a telephone link with n trunks. Calls arrive in a Poisson stream with rate λ; each call occupies one trunk (if available) for an interval of time distributed exponentially with parameter μ. There is no queue; any call which finds all the servers (trunks) busy departs immediately. D.G. Kendall's notation for a Markovian system with n servers and room for not more than n jobs is M/M/n/n.

The number of calls in progress (which is the same as the number of busy trunks) is a Birth-and-Death process with a finite state space, $\{0, 1, \ldots, n\}$. The arrival and departure rates are $\lambda_j = \lambda$, for $j = 0, 1, \ldots, n-1$, and $\mu_j = j\mu$, for $j = 1, 2, \ldots, n$. The steady-state probabilities, p_j, always exist. They follow the first pattern in (5.60):

$$p_j = \frac{\rho^j}{j!} p_0 \; ; \; j = 0, 1, \ldots, n , \qquad (5.70)$$

where $\rho = \lambda/\mu$. The normalizing equation then yields

$$p_0 = \left[\sum_{j=0}^{n} \frac{\rho^j}{j!} \right]^{-1} . \qquad (5.71)$$

The chief performance measure of the M/M/n/n model is the probability of losing a call, i.e. the probability of finding all lines busy, p_n. This is given by

$$p_n = \frac{\rho^n}{n!} \left[\sum_{j=0}^{n} \frac{\rho^j}{j!} \right]^{-1} . \qquad (5.72)$$

Expression (5.72) is known as 'Erlang's loss formula'.

The system throughput, T, defined as the average number of calls that are admitted (and completed) per unit time, is obtained from

$$T = \lambda(1 - p_n) . \qquad (5.73)$$

Example

2. A telephone engineer may be faced with the problem of determining the optimal number of trunks for a given load. Assuming, for simplicity, that the revenue derived from the link is proportional to the throughput, and the cost is proportional to the number of trunks, the problem becomes one of maximizing a non-linear objective function:

$$\max_{n \geq 1}(c_1 T - c_2 n),$$

where c_1 and c_2 are non-negative constants. This is equivalent to finding the minimum

$$\min_{n \geq 1}(\lambda c_1 p_n + c_2 n).$$

There is always an optimal value for n. Whether that value is 1, or is greater than 1, depends on the load ρ and on the ratio c_1/c_2.

* * *

Consider now a model with Poisson arrivals and an infinite number of servers. There is no queue, and no rejections: every incoming job starts service immediately. This is the M/M/∞ system; in the context of queueing networks we have also referred to it as the 'independent delay' system.

M/M/∞ can be regarded as the limiting case of either M/M/n or the Erlang model, as $n \to \infty$. The state space is infinite, $\{0, 1, \ldots\}$, and the arrival and departure rates are $\lambda_j = \lambda$, $\mu_j = j\mu$ for all j. The general solution of the balance equations is of the form

$$p_j = \frac{\rho^j}{j!} p_0 \ ; \ j = 0, 1, \ldots . \tag{5.74}$$

The normalization yields

$$p_0 = \left[\sum_{j=0}^{\infty} \frac{\rho^j}{j!}\right]^{-1} = e^{-\rho}. \tag{5.75}$$

Note that the series always converges, therefore a steady-state distribution always exists. This is not surprising, since the service capacity is infinite and no offered load can saturate it.

Thus the steady-state number of jobs in the system, which is the same as the number of busy servers, has the Poisson distribution with parameter ρ:

$$p_j = \frac{\rho^j}{j!} e^{-\rho} \; ; \; j = 0, 1, \ldots . \tag{5.76}$$

The mean and the variance of that number are equal to ρ.

5.5.3 The M/M/1/N queue

In practice, queues are rarely unbounded. Jobs are buffered, and buffers are of finite size. Perhaps the simplest model of this type is the M/M/1 queue with a limited waiting room. This was introduced as an example in section 5.2, but was not analysed there.

The assumptions of the M/M/1/N model are the same as for M/M/1, except that at most N jobs (including the one in service) are admitted into the system. Any job which finds N jobs present on arrival is lost.

This is a Birth-and-Death process on the state space $\{0, 1, \ldots, N\}$. The corresponding state diagram is shown in figure 5.8: the arrival rates are equal to $\lambda_j = \lambda$ $(j = 0, 1, \ldots, N-1)$, and the departure rates are $\mu_j = \mu$, $(j = 1, 2, \ldots, N)$. The solution of the balance equations is

$$p_j = \rho^j p_0 \; ; \; j = 0, 1, \ldots, N, \tag{5.77}$$

where $\rho = \lambda/\mu$. After normalization, this becomes

$$p_j = \frac{(1-\rho)\rho^j}{1-\rho^{N+1}} \; ; \; j = 0, 1, \ldots, N. \tag{5.78}$$

Note the effect of the offered load on the shape of that distribution. When $\rho < 1$, the probabilities p_j decrease with j; the most likely state is the empty system (in this case the unbounded M/M/1 queue is stable). When $\rho = 1$, all states are equally likely (the right-hand side of (5.78) reduces to $p_j = 1/(N+1)$, for all j). When $\rho > 1$, the probabilities p_j increase with j and the most likely state is the one where the queue is full. We also observe a predictable behaviour in the limit $N \to \infty$: (5.78) approaches the M/M/1 distribution when $\rho < 1$, and it approaches 0, for all j, when $\rho \geq 1$.

The steady-state average number of jobs in the M/M/1/N system is given by

$$L = \sum_{j=1}^{N} j p_j = \frac{\rho}{1-\rho} \frac{1 - (N+1)\rho^N + N\rho^{N+1}}{1 - \rho^{N+1}} . \tag{5.79}$$

This expression has the value $N/2$ when $\rho = 1$.

5.5 Birth-and-Death queueing models

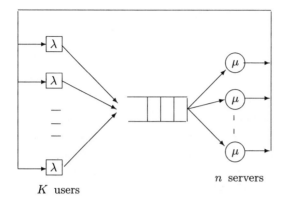

Fig. 5.14. The M/M/n/·/K model

The throughput, T, is obtained from

$$T = (1 - p_0)\mu = (1 - p_N)\lambda ; \qquad (5.80)$$

jobs depart at rate μ while the server is busy; alternatively, they join at rate λ while the queue is not full.

The average response time, W, of a job that is admitted into the system can be found from Little's theorem: $W = L/T$.

5.5.4 The M/M/n/·/K model

A general model of a system with K users and n parallel servers was described in section 3.1.1. A relation between the average response time and the throughput was derived (the 'response time law'), without showing how those quantities can be computed. Now we can fill that gap, under suitable assumptions.

The model is illustrated in figure 5.14. We can assume, without loss of generality, that there are at least as many users as servers (if $K < n$, then $n - K$ servers are never used and may be discarded). The user think times are distributed exponentially with parameter λ. Service times at all servers are distributed exponentially with parameter μ. The system is in state j ($j = 0, 1, \ldots, K$) if j users are waiting for their jobs to be completed and $K - j$ users are thinking.

This model is denoted M/M/n/·/K (the '·' indicates that there are no restrictions on the queue size, beyond the one implied by the number of

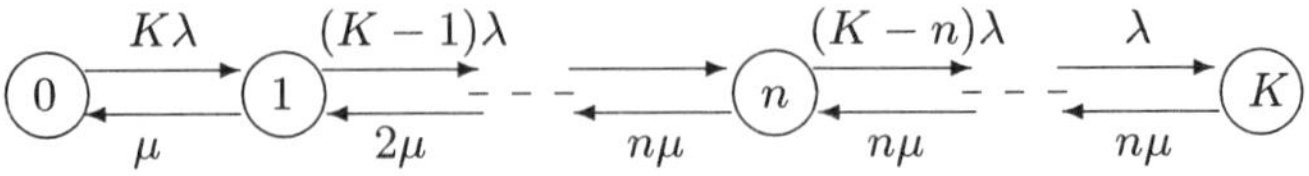

Fig. 5.15. State diagram for the M/M/n/·/K model

sources). The instantaneous transition rate from state j to state $j+1$ is equal to $\lambda_j = (K-j)\lambda$ ($j = 0, 1, \ldots, K-1$), since each of the thinking users submits jobs at rate λ. The rates from state j to state $j-1$ depend on whether the number of jobs is less than the number of servers, in a similar way to the M/M/n model:

$$\mu_j = \begin{cases} j\mu & \text{for } j = 1, 2, \ldots, n-1 \\ n\mu & \text{for } j = n, n+1, \ldots, K \end{cases}$$

The state diagram is shown in figure 5.15.

The balance and normalizing equations yield

$$p_j = \frac{K!}{(K-j)!j!} \rho^j p_0 \ ; \ j = 0, 1, \ldots, n-1$$

$$p_j = \frac{K!}{(K-j)!n!n^{j-n}} \rho^j p_0 \ ; \ j = n, n+1, \ldots, K, \quad (5.81)$$

with p_0 given by

$$p_0 = \left[\sum_{j=0}^{n-1} \frac{K!}{(K-j)!j!} \rho^j + \sum_{j=n}^{K} \frac{K!}{(K-j)!n!n^{j-n}} \rho^j \right]^{-1}. \quad (5.82)$$

The throughput, T, can be obtained either as the average number of job completions, or as the average number of job submissions, per unit time. The former approach requires the average number of busy servers, r:

$$r = \sum_{j=1}^{n-1} j p_j + n \sum_{j=n}^{K} p_j . \quad (5.83)$$

The expression for the throughput is then $T = r\mu$. Alternatively, we could find the average number of jobs in service or in the queue, L:

$$L = \sum_{j=1}^{K} j p_j . \quad (5.84)$$

Then the average number of thinking users is $K - L$. Since each of them submits jobs at rate λ, the throughput is equal to $T = (K - L)\lambda$.

5.5 Birth-and-Death queueing models

When T has been found, the response time law provides the average response time, W:

$$W = \frac{K}{T} - \frac{1}{\lambda}. \tag{5.85}$$

In the two special cases when $n = 1$ and $n = K$, the expressions have a simpler form. If there is a single server, the steady-state probabilities are

$$p_j = \frac{\rho^j}{(K-j)!} \left[\sum_{i=0}^{K} \frac{\rho^i}{(K-i)!} \right]^{-1} \quad ; \; j = 0, 1, \ldots, K, \tag{5.86}$$

and the throughput is equal to $T = (1 - p_0)\mu$.

When the number of servers is equal to the number of users, no job has to queue and users do not interfere with each other in any way. The steady-state distribution of the number of jobs in service is binomial:

$$p_j = \binom{K}{j} \left(\frac{\rho}{1+\rho} \right)^j \left(\frac{1}{1+\rho} \right)^{K-j} \quad ; \; j = 0, 1, \ldots, K. \tag{5.87}$$

The average number of busy servers is $r = K\rho/(1+\rho)$. The throughput is given by

$$T = \frac{K\lambda}{1+\rho}.$$

It was pointed out in section 3.1.1 that one of the applications of the M/M/$n/\cdot/K$ system is to model machine maintenance. In that context, the special case $n = K$ represents K machines which break down and are repaired independently of each other (there is a repairman for each machine). Expression (5.87) gives the distribution of broken machines or, if λ and μ are exchanged, the distribution of the operative machines.

Exercises

1. Consider an M/M/1 queue where the jobs' willingness to join the queue is influenced by the latter's size. More precisely, a job which finds j other jobs in the system joins the queue with probability $1/(j+1)$ and departs immediately with probability $j/(j+1)$ ($j = 0, 1, \ldots$).

Write and solve the balance equations for the limiting probabilities p_j. Show that a steady-state distribution always exists. Find the steady-state utilization of the server (the probability it is busy), the throughput (average number of jobs departing per unit time), the average number of

198   5 Markov chains and processes

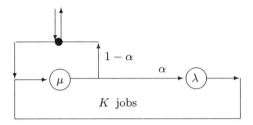

Fig. 5.16. A closed network with two single-server nodes

jobs in the system and the average response time for a job that decides to join.

2. A computer system consists of five parallel processors, whose primary purpose is to serve user jobs. The latter arrive in a Poisson stream at the rate of 20 jobs per minute; job sevice times are distributed exponentially with mean 9 seconds. Whenever a processor has no user jobs to run, it is assigned to secondary low-priority tasks, of which there is always an unlimited supply. The average execution time for a secondary task is 5 seconds. If a user job arrives and needs a processor which is running a secondary task, the latter is interrupted and eventually resumed when a processor is again available.

Using the M/M/n results, find (a) the average number of servers busy with user jobs and hence the service capacity available for the secondary tasks, (b) the average number of secondary jobs that are completed per minute, (c) the probability that a user job has to wait, (d) the average waiting time for a user job.

3. A fixed number of jobs, K, circulate between two service nodes, each with its own queue and a processor. After completing service at node 1, a job goes to node 2 with probability α and leaves the system with probability $1 - \alpha$; in the latter case, the job is immediately replaced by a new job at node 1. After completing service at node 2, jobs return to node 1. Service times at nodes 1 and 2 are exponentially distributed, with parameters μ and λ, respectively. This simple closed queueing network is illustrated in figure 5.16.

Let p_j be the steady-state probability that there are j jobs at node 1 (waiting and/or in service); $j = 0, 1, \ldots, K$. Show, by writing and solving the balance and normalizing equations, that p_j is equal to the probability that there are j jobs in an M/M/1/K system with arrival rate λ and service time parameter $\alpha\mu$. Find the throughput of the present system, i.e. the average number of departures per unit time.

4. A company providing networked services has a single central processor and K users. It can be modelled as an M/M/1/·/K system. The users pay in proportion to their demand for service: the charge is c per unit of processor time used.

What is the average revenue received by the company per unit time? One fine day, the managing director decides to double the charge. As a result, half of the users disconnect their computers and transfer their custom elsewhere (assume that K is even). Under what conditions, if any, is that move profitable?

5. Modify the M/M/n/·/K model by assuming that there is no room for a queue: any job which is submitted when all servers are busy is discarded and the corresponding user enters a new think period. Solve the balance equations and find the distribution of the number of jobs in service (this is known as the 'Engset distribution'). Hence derive expressions for the throughput (average number of successfully completed jobs per unit time) and the average number of discarded jobs per unit time.

6. Generalize the model from example 3 in section 5.3 by considering a telephone network with K subscribers (assume that K is even). Find the steady-state probability, p_j, that there are j calls in progress ($j = 0, 1, \ldots, K/2$) and the average number of calls lost per unit time.

5.6 Literature

Our treatment of Markov chains and Markov processes is similar to that of Çinlar [1]. The interested reader is directed to that book for additional information and for the rigorous proofs of the fundamental steady-state theorems. The spectral method for solving Kolmogorov's differential equations is discussed in an appendix. Among the many other books on stochastic processes which may be consulted are Doob [2] and Parzen [6].

Queuing applications to telephony appeared as early as 1917, in the work of Erlang [4], Engset [3], and other pioneers. The embedded Markov chain analysis of the M/G/1 queue was published in [5]. The literature on

Markov models and queues is now very extensive. Some of the references in this area were mentioned at the end of chapter 3.

References

1. E. Çinlar, *Introduction to Stochastic Processes*, Prentice-Hall, 1975.
2. J.L. Doob, *Stochastic Processes*, John Wiley, 1953.
3. T. Engset, "Emploi du calcul des probabilités pour la détermination du nombre de sélecteurs dans les Bureaux Téléphoniques Centraux", *Rev. gén. elect.*, **9**, 138–140, 1921.
4. A.K. Erlang, "Solution of Some Probability Problems of Significance for Automatic Telephone Exchanges", *Electroteknikeren*, **13**, 5–13, 1917.
5. D.G. Kendall, "Stochastic Processes Occurring in the Theory of Queues and Their Analysis by the Method of the Embedded Markov Chain", *Annals of Mathematical Statistics*, **24**, 338–354, 1953.
6. E. Parzen, *Stochastic Processes*, Holden-Day, 1962.

6

Queues in Markovian environments

We shall consider a large class of queueing models where the arrival and/or service mechanisms are influenced by some external processes. A single unbounded queue evolves in an environment which changes state from time to time. The instantaneous job arrival rate and the instantaneous service rate may depend on the state of the environment and also, to a limited extent, on the number of jobs.

The system state at time t is described by a pair of random variables, (X_t, Y_t), where X_t is the state of the environment and Y_t is the number of jobs present. The variable X_t takes a finite number of values, numbered $0, 1, \ldots, N$; these are also called the 'operational modes' of the service facility. The possible values of Y_t are, as always, $0, 1, \ldots$. Thus, the system is in state (i, j) when the operational mode is i and there are j jobs waiting and/or being served.

The two-dimensional process $X = \{(X_t, Y_t) : t \geq 0\}$ is assumed to have the Markov property, i.e. given the current operational mode and number of jobs, the future behaviour of X is independent of its past history. That process is said to model a 'queue in Markovian environment'. The state space of X, $\{0, 1, \ldots, N\} \times \{0, 1, \ldots\}$, is a 'lattice strip'.

The evolution of the process X is governed by the instantaneous transition rates from state (i, j) to state (i', j'). These rates may depend on the operational modes, i and i', in an arbitrary way. However, their dependence on the number of jobs is restricted:

- There is a threshold M, such that the instantaneous transition rates out of state (i, j) do not depend on j when $j \geq M$.

In addition to the above restriction, we shall be interested primarily in queueing models where jobs arrive and depart singly. That is, when Y_t changes value, it does so only from j to $j+1$ or to $j-1$. Such processes are said to be of the *Quasi-Birth-and-Death* type (the term *skip-free* is also used).

6.1 Quasi-Birth-and-Death models

The process $X = \{(X_t, Y_t) : t \geq 0\}$ evolves according to the following set of instantaneous transition rates:

(a) From state (i,j) to state (k,j) $(0 \leq i, k \leq N\ ;\ i \neq k\ ;\ j \geq 0)$, with rate $a_j(i,k)$;

(b) From state (i,j) to state $(k, j+1)$ $(0 \leq i, k \leq N\ ;\ j \geq 0)$, with rate $b_j(i,k)$;

(c) From state (i,j) to state $(k, j-1)$ $(0 \leq i, k \leq N\ ;\ j \geq 1)$, with rate $c_j(i,k)$.

It is convenient to define the matrices associated with the rates (a), (b) and (c): $A_j = [a_j(i,k)]$, $B_j = [b_j(i,k)]$ and $C_j = [c_j(i,k)]$, respectively (the main diagonal of A_j is zero by definition; also, $C_0 = 0$ by definition). There is a threshold, M ($M \geq 1$), such that those matrices do not depend on j when $j \geq M$. In other words,

$$A_j = A\ ;\ B_j = B\ ;\ C_j = C\ ;\ j \geq M\ . \qquad (6.1)$$

Transitions (a) correspond to changes in the operational mode. Those of type (b) represent a job arrival coinciding with a change in the operational mode. If arrivals are not accompanied by such changes, then the matrices B_j and B are diagonal. Similarly, a transition of type (c) represents a job departure coinciding with a change in the operational mode. Again, if such coincidences do not occur, then the matrices C and C_j are diagonal. The shape of the state diagram of a Quasi-Birth-and-Death (QBD) process, showing some transitions out of state (i,j), is presented in figure 6.1.

The requirement that all transition rates cease to depend on the size of the job queue beyond a certain threshold is not too restrictive. It allows the consideration of models where the arrival, service and mode transition rates depend on the current operational mode. On the other hand, it is difficult to think of applications where those rates would depend on the number of jobs waiting for service. Note that we impose no limit on the magnitude of the threshold M, although it must be pointed out that the larger M is, the greater the complexity of the solution.

6.1 Quasi-Birth-and-Death models

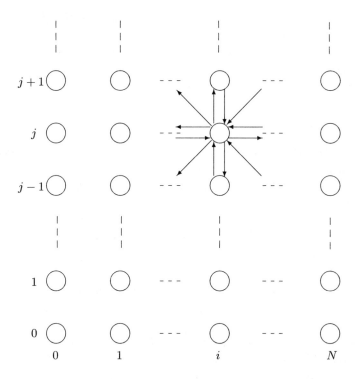

Fig. 6.1. State diagram of a QBD process

The object of the analysis is to determine the joint steady-state distribution of the operational mode and the number of jobs in the system:

$$p_{i,j} = \lim_{t \to \infty} P(X_t = i, Y_t = j) \; ; \; i = 0, 1, \ldots, N \; ; \; j = 0, 1, \ldots . \quad (6.2)$$

That distribution exists for an irreducible Markov process if, and only if, the corresponding set of balance equations has a positive solution that can be normalized.

The marginal distributions of the number of jobs in the system, and of the operational mode, can be obtained from the joint distribution:

$$p_{\cdot,j} = \sum_{i=0}^{N} p_{i,j} . \quad (6.3)$$

$$p_{i,\cdot} = \sum_{j=0}^{\infty} p_{i,j} . \quad (6.4)$$

Various performance measures can be obtained in terms of these joint and marginal distributions.

Before addressing the problem of how to solve the model and determine the probabilities (6.2), it would be instructive to show some examples of systems giving rise to QBD processes.

6.1.1 The MMPP/M/1 queue

The 'Markov-Modulated Poisson Process' (MMPP) offers a more general mechanism for modelling arrivals than the Poisson stream. The arrival rate is not fixed, but is controlled by an irreducible Markov process (environment), X_t, with $N+1$ states numbered $0, 1, \ldots, N$. If the environment is in state i at time t, then there is an arrival in a small interval $(t, t+h)$ with probability $\lambda_i h + o(h)$ (i.e. when $X_t = i$, the instantaneous arrival rate is λ_i). The Poisson process is of course a special case ($N = 0$) of MMPP.

Consider a single-server queueing system with exponentially distributed service times (parameter μ), and MMPP arrivals. The above assumptions ensure that the pair $X = \{(X_t, Y_t) : t \geq 0\}$ (Y_t being the number of jobs in the system at time t) is a QBD process. The j-independence threshold, M, is 0 in this case.

Denote the generator matrix of X_t by $\tilde{A}$. The matrix A, defined in (6.1), containing the instantaneous transition rates for changes in X_t only, is obtained from $\tilde{A}$ by replacing the latter's main diagonal with 0. When the system is in state (i, j), jobs arrive at rate λ_i and depart at rate μ ($j > 0$). Hence, the matrices B and C are diagonal and are given by

$$B = \begin{bmatrix} \lambda_0 & & & \\ & \lambda_1 & & \\ & & \ddots & \\ & & & \lambda_N \end{bmatrix} \; ; \; C = \begin{bmatrix} \mu & & & \\ & \mu & & \\ & & \ddots & \\ & & & \mu \end{bmatrix}. \quad (6.5)$$

In this example, since the generator matrix $\tilde{A}$ is irreducible and finite, the marginal steady-state distribution of X_t always exists. Moreover, since $\tilde{A}$ does not depend on the queue size, that distribution can be determined separately. The row vector $\boldsymbol{\pi} = (p_{0,\cdot}, p_{1,\cdot}, \ldots, p_{N,\cdot})$ satisfies the balance and normalizing equations

$$\boldsymbol{\pi}\tilde{A} = 0 \; ; \; \sum_{i=0}^{N} p_{i,\cdot} = 1 \, . \quad (6.6)$$

6.1 Quasi-Birth-and-Death models

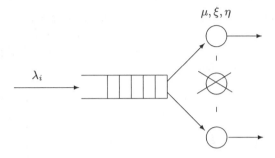

Fig. 6.2. A multiserver queue with breakdowns and repairs

The steady-state average arrival rate, λ, is equal to

$$\lambda = \sum_{i=0}^{N} p_i \cdot \lambda_i . \qquad (6.7)$$

Intuitively, the queue is stable if, and only if, the offered load, λ/μ, is less than 1. This can be proved rigorously.

6.1.2 A multiserver queue with breakdowns and repairs

A single, unbounded queue is served by N identical parallel servers. Each server goes through alternating periods of being operative and inoperative, independently of the others and of the number of jobs in the system. The operative and inoperative periods are distributed exponentially with parameters ξ and η, respectively. Thus, the number of operative servers at time t, X_t, is a Markov process on the state space $\{0, 1, \ldots, N\}$.

Jobs arrive according to a Markov-Modulated Poisson Process controlled by X_t. When there are i operative servers, the instantaneous arrival rate is λ_i. Jobs are taken for service from the front of the queue, one at a time, by available operative servers. The required service times are distributed exponentially with parameter μ. An operative server cannot be idle if there are jobs waiting to be served. A job whose service is interrupted by a server breakdown is returned to the front of the queue. When an operative server becomes available, the service is resumed from the point of interruption, without any switching overheads. The flow of jobs is shown in figure 6.2.

This system is modelled by the QBD process $X = \{(X_t, Y_t) : t \geq 0\}$, where X_t is the number of servers that are operative at time t, and Y_t

is the number of jobs present. The instantaneous transition rates out of state (i, j) that are of type (a) (corresponding to changes in X_t only) do not depend on j: one of the i operative servers breaks down with rate $i\xi$, and one of the $(N-i)$ broken servers is repaired with rate $(N-i)\eta$. The matrices A_j are given by:

$$A_j = A = \begin{bmatrix} 0 & N\eta & & & \\ \xi & 0 & (N-1)\eta & & \\ & 2\xi & 0 & \ddots & \\ & & \ddots & \ddots & \eta \\ & & & N\xi & 0 \end{bmatrix}.$$

The transition rates of type (b) (corresponding to arrivals) also do not depend on j. The matrix B is the same as in (6.5).

On the other hand, the departure rates depend both on the number of operative servers and the number of jobs present. If the system is in state (i, j) with $i > j$, then there are j services in progress and the departure rate is $j\mu$; if $i \leq j$, the departure rate is $i\mu$. Denote

$$\mu_{i,j} = \min(i, j)\mu \; ; \; i = 0, 1, \ldots, N \; ; \; j = 1, 2, \ldots \; .$$

The matrices C_j, containing the instantaneous transition rates corresponding to departures, can thus be written as

$$C_j = \begin{bmatrix} \mu_{0,j} & & & \\ & \mu_{1,j} & & \\ & & \ddots & \\ & & & \mu_{N,j} \end{bmatrix} \; ; \; j = 1, 2, \ldots \; ,$$

where, in all cases, $\mu_{0,j} = 0$ and $\mu_{1,j} = \mu$. These matrices cease to depend on j when $j \geq N$. Thus, the threshold M is now equal to N.

The marginal distribution of the number of operative servers is binomial (see section 5.5.4):

$$p_{i,\cdot} = \binom{N}{i} \left(\frac{\eta}{\xi+\eta}\right)^i \left(\frac{\xi}{\xi+\eta}\right)^{N-i} \; ; \; i = 0, 1, \ldots, N \; . \qquad (6.8)$$

Hence, the steady-state average number of operative servers is equal to

$$E(X_t) = \frac{N\eta}{\xi+\eta} \; . \qquad (6.9)$$

6.1 Quasi-Birth-and-Death models

Fig. 6.3. Two nodes with a finite intermediate buffer

As in the previous example, the overall average arrival rate is

$$\lambda = \sum_{i=0}^{N} p_{i,\cdot} \lambda_i . \qquad (6.10)$$

The condition for stability of X is that the offered load is less than the processing capacity:

$$\frac{\lambda}{\mu} < \frac{N\eta}{\xi + \eta} . \qquad (6.11)$$

6.1.3 Manufacturing blocking

Consider a network of two nodes in tandem, such as the one in figure 6.3. Jobs arrive into the first node in a Poisson stream with rate λ, and join an unbounded queue. After completing service at node 1 (exponentially distributed with parameter μ), they attempt to go to node 2, where there is a finite buffer with room for a maximum of $N-1$ jobs (including the one in service). If that transfer is impossible because the buffer is full, the job remains at node 1, preventing its server from starting a new service, until the completion of the current service at node 2 (exponentially distributed with parameter ξ). In this last case, server 1 is said to be 'blocked'. Transfers from node 1 to node 2 are instantaneous.

The above type of blocking is referred to as 'manufacturing blocking', to distinguish it from 'communication blocking', where node 1 checks the buffer before starting a service and if full does not start.

In this system, the unbounded queue at node 1 operates in a finite-state environment defined by node 2. We say that the environment, X_t, is in state i if there are i jobs at node 2 and server 1 is not blocked ($i = 0, 1, \ldots, N-1$). An extra state, $X_t = N$, is needed to describe the situation where there are $N-1$ jobs at node 2 and server 1 is blocked. Then the pair $X = \{(X_t, Y_t) : t \geq 0\}$, where Y_t is the number of jobs at node 1, is a QBD process.

If the system is in state (i, j), with $0 < i \leq N-1$, it can move to state $(i-1, j)$ with rate ξ. From state $(N-1, j)$, for $j > 0$, it can also move to state (N, j) with rate μ. From state (N, j) the environment can change without changing the node 1 queue only if $j = 0$; then the transition is to state $(N - 1, 0)$ with rate ξ. Thus, the environmental transition rate matrices, A_0 and $A_j = A$ $(j > 0)$, are given by

$$A_0 = \begin{bmatrix} 0 & & & & \\ \xi & 0 & & & \\ & \ddots & \ddots & & \\ & & \xi & 0 & \\ & & & 0 & 0 \end{bmatrix} \quad ; \quad A = \begin{bmatrix} 0 & 0 & & & & \\ \xi & 0 & 0 & & & \\ & \ddots & \ddots & \ddots & & \\ & & \xi & 0 & \mu \\ & & & 0 & 0 \end{bmatrix}.$$

The last row of A_0 is 0 by definition; the state $(N, 0)$ is in fact unreachable.

Since the arrival rate into node 1 does not depend on either i or j, we have $B_j = B = \lambda I$, where I is the identity matrix of order $N + 1$. The departures from node 1 (which can occur when $i \neq N - 1$) are always accompanied by environmental changes: from state (i, j) the system moves to state $(i + 1, j - 1)$ with rate μ for $i < N - 1$; from state (N, j) to state $(N - 2, j - 1)$ with rate ξ. Hence, the departure rate matrices do not depend on j and are equal to

$$C_j = C = \begin{bmatrix} 0 & \mu & 0 & & & \\ 0 & 0 & \mu & & & \\ & & \ddots & \ddots & & \\ & & & 0 & \mu & 0 \\ & & & 0 & 0 & 0 \\ & & & \xi & 0 & 0 \end{bmatrix}.$$

In this example, the j-independency threshold is $M = 1$. Because of the difference between A_0 and A, the marginal distribution of the number of jobs at node 2 cannot be determined without finding the joint distribution of X_t and Y_t.

6.1.4 Phase-type distributions

There is a large and useful family of distributions that can be incorporated into queueing models by means of Markovian environments. Those distributions are 'almost' general, in the sense that any distribution function either belongs to this family or can be approximated as closely as desired by functions from it.

6.1 Quasi-Birth-and-Death models

Let X_t be a Markov process with state space $\{0, 1, \ldots, N\}$ and generator matrix $\tilde{A}$. States $0, 1, \ldots, N-1$ are transient, while state N, reachable from any of the other states, is absorbing (the last row of $\tilde{A}$ is 0). At time 0, the process starts in state i with probability α_i ($i = 0, 1, \ldots, N-1$; $\alpha_0 + \alpha_1 + \ldots + \alpha_{N-1} = 1$). Eventually, after an interval of length T, it is absorbed in state N. The random variable T is said to have a 'phase-type' (PH) distribution with parameters $\tilde{A}$ and α_i.

The exponential distribution is obviously phase-type ($N = 1$). So is the Erlang distribution—the convolution of N exponentials (exercise 5 in section 2.3). The corresponding generator matrix is

$$\tilde{A} = \begin{bmatrix} -\mu & \mu & & & \\ & -\mu & \mu & & \\ & & \ddots & \ddots & \\ & & & -\mu & \mu \\ & & & & 0 \end{bmatrix},$$

and the initial probabilities are $\alpha_0 = 1$, $\alpha_1 = \ldots = \alpha_{N-1} = 0$.

Another common PH distribution is the 'hyperexponential', where $X_0 = i$ with probability α_i, and absorption occurs at the first transition. The generator matrix of the hyperexponential distribution is

$$\tilde{A} = \begin{bmatrix} -\mu_0 & & & & \mu_0 \\ & -\mu_1 & & & \mu_1 \\ & & \ddots & & \vdots \\ & & & -\mu_{N-1} & \mu_{N-1} \\ & & & & 0 \end{bmatrix}.$$

The corresponding probability distribution function, $F(x)$, is a mixture of exponentials:

$$F(x) = 1 - \sum_{i=0}^{N-1} \alpha_i e^{-\mu_i x}.$$

The PH family is very versatile. It contains distributions with both low and high coefficients of variation. It is closed with respect to mixing and convolution: if X_1 and X_2 are two independent PH random variables with N_1 and N_2 (non-absorbing) phases respectively, and c_1 and c_2 are constants, then $c_1 X_1 + c_2 X_2$ has a PH distribution with $N_1 + N_2$ phases.

A model with a single unbounded queue, where either the interarrival intervals, or the service times, or both, have PH distributions, is easily cast in the framework of a queue in Markovian environment. Consider, for instance, the M/PH/1 queue. Its state at time t can be represented as a pair (X_t, Y_t), where Y_t is the number of jobs present and X_t is the phase of the current service (if $Y_t > 0$). When X_t has a transition into the absorbing state, the current service completes and (if the queue is not empty) a new service starts immediately, entering phase i with probability α_i.

The PH/PH/n queue can also be represented as a QBD process. However, the state of the environmental variable, X_t, now has to indicate the phase of the current interarrival interval and the phases of the current services at all busy servers. If the interarrival interval has N_1 phases and the service has N_2 phases, the state space of X_t will be of size $N_1 N_2^n$.

6.2 Solution methods

Let us now return to the problem of determining the steady-state joint distribution of a QBD process, defined in (6.2). The probabilities $p_{i,j}$ satisfy the following set of balance equations:

$$p_{i,j} \sum_{k=0}^{N} [a_j(i,k) + b_j(i,k) + c_j(i,k)]$$

$$= \sum_{k=0}^{N} [p_{k,j} a_j(k,i) + p_{k,j-1} b_{j-1}(k,i) + p_{k,j+1} c_{j+1}(k,i)], \quad (6.12)$$

where $p_{i,-1} = b_{-1}(k,i) = c_0(i,k) = 0$ by definition (the left-hand side gives the total average number of transitions out of state (i,j) per unit time, while the right-hand side expresses the total average number of transitions into it). These balance equations can be written more compactly by introducing the row vectors of probabilities corresponding to states with j jobs in the system:

$$\mathbf{v}_j = (p_{0,j}, p_{1,j}, \ldots, p_{N,j}) \; ; \; j = 0, 1, \ldots . \quad (6.13)$$

Also, let D_j^A, D_j^B and D_j^C be the diagonal matrices whose ith diagonal element is equal to the ith row sum of A_j, B_j and C_j, respectively. Then equations (6.12), for $j = 0, 1, \ldots$, can be written as

$$\mathbf{v}_j [D_j^A + D_j^B + D_j^C] = \mathbf{v}_{j-1} B_{j-1} + \mathbf{v}_j A_j + \mathbf{v}_{j+1} C_{j+1}, \quad (6.14)$$

where $\mathbf{v}_{j-1} = \mathbf{0}$ and $D_0^C = B_{-1} = 0$ by definition.

6.2 Solution methods

When j is greater than the threshold M, the coefficients in (6.14) cease to depend on j:

$$\mathbf{v}_j[D^A + D^B + D^C] = \mathbf{v}_{j-1}B + \mathbf{v}_j A + \mathbf{v}_{j+1}C , \qquad (6.15)$$

for $j = M+1, M+2, \ldots$.

In addition, all probabilities must sum up to 1:

$$\sum_{j=0}^{\infty} \mathbf{v}_j \mathbf{e} = 1 , \qquad (6.16)$$

where $\mathbf{e}$ is a column vector with $N+1$ elements, all of which are equal to 1.

The first step of any solution method is to find the general solution of the infinite set of balance equations with constant coefficients, (6.15). The latter are normally written in the form of a homogeneous vector difference equation of order 2:

$$\mathbf{v}_j Q_0 + \mathbf{v}_{j+1} Q_1 + \mathbf{v}_{j+2} Q_2 = \mathbf{0} \; ; \; j = M, M+1, \ldots , \qquad (6.17)$$

where $Q_0 = B$, $Q_1 = A - D^A - D^B - D^C$ and $Q_2 = C$. There is more than one way of solving such equations.

6.2.1 Spectral expansion method

Associated with equation (6.17) is the so-called 'characteristic matrix polynomial', $Q(x)$, defined as

$$Q(x) = Q_0 + Q_1 x + Q_2 x^2 . \qquad (6.18)$$

Denote by x_k and $\mathbf{u}_k$ the 'generalized eigenvalues', and corresponding 'generalized left eigenvectors', of $Q(x)$. In other words, these are quantities which satisfy

$$\det[Q(x_k)] = 0 ,$$

$$\mathbf{u}_k Q(x_k) = \mathbf{0} \; ; \; k = 1, 2, \ldots, d , \qquad (6.19)$$

where $\det[Q(x)]$ is the determinant of $Q(x)$ and d is its degree. In what follows, the qualification *generalized* will be omitted.

The above eigenvalues do not have to be simple, but it is assumed that if one of them has multiplicity m, then it also has m linearly independent left eigenvectors. This tends to be the case in practice. So, the numbering in (6.19) is such that each eigenvalue is counted according to its multiplicity.

It is readily seen that if x_k and $\mathbf{u}_k$ are any eigenvalue and corresponding left eigenvector, then the sequence

$$\mathbf{v}_{k,j} = \mathbf{u}_k x_k^{j-M} \; ; \; j = M, M+1, \ldots , \qquad (6.20)$$

is a solution of equation (6.17). Indeed, substituting (6.20) into (6.17) we get

$$\mathbf{v}_{k,j} Q_0 + \mathbf{v}_{k,j+1} Q_1 + \mathbf{v}_{k,j+2} Q_2 = x_k^{j-M} \mathbf{u}_k [Q_0 + Q_1 x_k + Q_2 x_k^2] = \mathbf{0} .$$

By combining any multiple eigenvalues with each of their independent eigenvectors, we thus obtain d linearly independent solutions of (6.17). On the other hand, it is known that there cannot be more than d linearly independent solutions. Therefore, any solution of (6.17) can be expressed as a linear combination of the d solutions (6.20):

$$\mathbf{v}_j = \sum_{k=1}^{d} \alpha_k \mathbf{u}_k x_k^{j-M} \; ; \; j = M, M+1, \ldots , \qquad (6.21)$$

where α_k ($k = 1, 2, \ldots, d$), are arbitrary (complex) constants.

However, the only solutions that are of interest in the present context are those which can be normalized to become probability distributions. Hence, it is necessary to select from the set (6.21) those sequences for which the series $\sum \mathbf{v}_j \mathbf{e}$ converges. This requirement implies that if $|x_k| \geq 1$ for some k, then the corresponding coefficient α_k must be 0.

So, suppose that c of the eigenvalues of $Q(x)$ are strictly inside the unit disk (each counted according to its multiplicity), while the others are on the circumference or outside. Order them so that $|x_k| < 1$ for $k = 1, 2, \ldots, c$. The corresponding independent eigenvectors are $\mathbf{u}_1, \mathbf{u}_2, \ldots, \mathbf{u}_c$. Then any normalizable solution of equation (6.17) can be expressed as

$$\mathbf{v}_j = \sum_{k=1}^{c} \alpha_k \mathbf{u}_k x_k^{j-M} \; ; \; j = M, M+1, \ldots , \qquad (6.22)$$

where α_k ($k = 1, 2, \ldots, c$), are some constants.

Expression (6.22) is referred to as the 'spectral expansion' of the vectors $\mathbf{v}_j$. The coefficients of that expansion, α_k, are yet to be determined.

Note that if there are non-real eigenvalues in the unit disk, then they appear in complex-conjugate pairs. The corresponding eigenvectors are also complex-conjugate. The same must be true for the appropriate pairs of constants α_k, in order that the right-hand side of (6.22) may be real. To ensure that it is also positive, the real parts of x_k, $\mathbf{u}_k$ and α_k should be positive.

6.2 Solution methods

So far we have obtained expressions for the vectors $\mathbf{v}_M, \mathbf{v}_{M+1}, \ldots$; these contain c unknown constants. Now it is time to consider the balance equations (6.14), for $j = 0, 1, \ldots, M$. This is a set of $(M+1)(N+1)$ linear equations with $M(N+1)$ unknown probabilities (the vectors $\mathbf{v}_j$ for $j = 0, 1, \ldots, M-1$), plus the c constants α_k. However, only $(M+1)(N+1) - 1$ of these equations are linearly independent, since the generator matrix of the Markov process is singular. On the other hand, an additional independent equation is provided by (6.16).

In order that this set of linearly independent equations has a unique solution, the number of unknowns must be equal to the number of equations, i.e. $(M+1)(N+1) = M(N+1) + c$, or $c = N+1$. This observation implies that the QBD process has a steady-state distribution if, and only if, the number of eigenvalues of $Q(x)$ strictly inside the unit disk is equal to the number of states of the Markovian environment.

In summary, the spectral expansion solution procedure consists of the following steps:

1. Compute the eigenvalues x_k and the corresponding left eigenvectors $\mathbf{u}_k$, of $Q(x)$. If $c \neq N+1$, then stop; a steady-state distribution does not exist.

2. Solve the finite set of linear equations (6.14), for $j = 0, 1, \ldots, M$, and (6.16), with $\mathbf{v}_M$ and $\mathbf{v}_{M+1}$ given by (6.22), to determine the constants α_k and the vectors $\mathbf{v}_j$ for $j < M$.

3. Use the obtained solution in order to determine various moments, marginal probabilities, percentiles and other system performance measures that may be of interest.

Careful attention should be paid to step 1. The 'brute force' approach which relies on first evaluating the scalar polynomial $\det[Q(x)]$, then finding its roots, is very inefficient for large N and is therefore not recommended. An alternative which is preferable in most cases is to reduce the quadratic eigenvalue–eigenvector problem

$$\mathbf{u}[Q_0 + Q_1 x + Q_2 x^2] = \mathbf{0} \tag{6.23}$$

to a linear one of the form $\mathbf{u}Q = x\mathbf{u}$, where Q is a matrix whose dimensions are twice as large as those of Q_0, Q_1 and Q_2. The latter problem is normally solved by applying various transformation techniques. Efficient routines for that purpose are available in most numerical packages.

This linearization can be achieved quite easily if the matrix $C = Q_2$ is non-singular. Indeed, after multiplying (6.23) on the right by Q_2^{-1}, it

becomes
$$\mathbf{u}[H_0 + H_1 x + I x^2] = \mathbf{0}, \tag{6.24}$$

where $H_0 = Q_0 C^{-1}$, $H_1 = Q_1 C^{-1}$, and I is the identity matrix. By introducing the vector $\mathbf{y} = x\mathbf{u}$, equation (6.24) can be rewritten in the equivalent linear form

$$[\mathbf{u}, \mathbf{y}] \begin{bmatrix} 0 & -H_0 \\ I & -H_1 \end{bmatrix} = x[\mathbf{u}, \mathbf{y}]. \tag{6.25}$$

If C is singular but B is not, a similar linearization is achieved by multiplying (6.23) on the right by B^{-1} and making a change of variable $x \to 1/x$. Then the relevant eigenvalues are those outside the unit disk.

If both B and C are singular, then the desired result is achieved by first making a change of variable, $x \to (\gamma + x)/(\gamma - x)$, where the value of γ is chosen so that the matrix $S = \gamma^2 Q_2 + \gamma Q_1 + Q_0$ is non-singular. In other words, γ can have any value which is not an eigenvalue of $Q(x)$. After that change of variable, multiplying the resulting equation by S^{-1} on the right reduces it to the form (6.24).

The computational demands of step 2 may be high if the threshold M is large. However, if the matrices B_j ($j = 0, 1, \ldots, M-1$) are non-singular (which is often the case in practice), then the vectors $\mathbf{v}_{M-1}, \mathbf{v}_{M-2}, \ldots, \mathbf{v}_0$ can be expressed in terms of $\mathbf{v}_M$ and $\mathbf{v}_{M+1}$, with the aid of equations (6.14) for $j = M, M-1, \ldots, 1$. One is then left with equations (6.14) for $j = 0$, plus (6.16) (a total of $N+1$ independent linear equations), for the $N+1$ unknowns x_k.

Having determined the coefficients in the expansion (6.22) and the probabilities $p_{i,j}$ for $j < N$, it is easy to compute performance measures. The steady-state probability that the environment is in state i is given by

$$p_{i,\cdot} = \sum_{j=0}^{N-1} p_{i,j} + \sum_{k=1}^{N+1} \alpha_k u_{k,i} \frac{1}{1 - x_k}, \tag{6.26}$$

where $u_{k,i}$ is the ith element of $\mathbf{u}_k$.

The conditional average number of jobs in the system, L_i, given that the environment is in state i, is obtained from

$$L_i = \frac{1}{p_{i,\cdot}} \left[\sum_{j=1}^{N-1} j p_{i,j} + \sum_{k=1}^{N+1} \alpha_k u_{k,i} \frac{M - (M-1) x_k}{(1 - x_k)^2} \right]. \tag{6.27}$$

6.2 Solution methods

The overall average number of jobs in the system, L, is equal to

$$L = \sum_{i=0}^{N} p_{i,.} L_i \ . \tag{6.28}$$

The spectral expansion solution can also be used to provide simple estimates of performance when the system is heavily loaded. The important observation in this connection is that when the system approaches instability, the expansion (6.22) is dominated by the eigenvalue with the largest modulus inside the unit disk, x_{N+1}. That eigenvalue is always real. It can be shown that when the offered load is high, the average number of jobs in the system is approximately equal to $x_{N+1}/(1 - x_{N+1})$.

6.2.2 Matrix-geometric solution

Consider the quadratic matrix equation whose coefficients are those of the difference equation (6.17):

$$Q_0 + RQ_1 + R^2 Q_2 = 0 \ . \tag{6.29}$$

If the matrix R is a solution of (6.29), then the sequence

$$\mathbf{v}_j = \mathbf{v} R^{j-M} \ ; \ j = M, M+1, \ldots \ , \tag{6.30}$$

where $\mathbf{v}$ is an arbitrary vector, is a solution of (6.17). This is immediately seen by substituting (6.30) into (6.17):

$$\mathbf{v}_j Q_0 + \mathbf{v}_{j+1} Q_1 + \mathbf{v}_{j+2} Q_2 = \mathbf{v} R^{j-M} [Q_0 + RQ_1 + R^2 Q_2] = \mathbf{0} \ .$$

The sequence (6.30) is known as the 'matrix-geometric' solution of the state-independent balance equations (6.17).

Remark. Equations (6.29) and (6.19) are closely related. Indeed, suppose that x and $\mathbf{u}$ are an eigenvalue and a left eigenvector of the matrix R, i.e. they satisfy $\mathbf{u}R = x\mathbf{u}$. It is readily verified that if R satisfies equation (6.29), then x and $\mathbf{u}$ satisfy (6.19). Conversely, if all eigenvalues and eigenvectors of R satisfy (6.19), then R satisfies (6.29).

In order that the matrix-geometric solution can be normalized to become a probability distribution, the matrix R must be non-negative and all its $N+1$ eigenvalues must be less than 1 in modulus. When those conditions are satisfied, the vector $\mathbf{v}$, appearing in the right-hand side of (6.30), is in fact the unknown probability vector $\mathbf{v}_N$. That vector, together with the remaining unknown probabilities, is determined from the balance equations (6.14) for $j \leq M$, and the normalizing equation. This part of the solution is very similar to step 2 of the spectral expansion algorithm, and has the same order of complexity.

Thus, the matrix-geometric solution of a QBD process consists of the following steps:

1. Find the minimal non-negative solution, R, of the quadratic matrix equation (6.29). If the spectral radius of R is greater than or equal to 1, then stop; a steady-state distribution does not exist.
2. Solve the finite set of linear equations (6.14), for $j = 0, 1, \ldots, M$ (with $\mathbf{v}_{M+1} = \mathbf{v}_M R$), together with (6.16), to determine the the vectors $\mathbf{v}_j$ for $j \leq M$.
3. Use the obtained solution to determine system performance measures.

The matrix equation (6.29) is usually solved by iteration. For instance, if Q_1 is non-singular, one can use the iterative scheme

$$R_{n+1} = [-Q_0 - R_n^2 Q_2] Q_1^{-1}, \qquad (6.31)$$

starting with $R_0 = 0$. Clearly, if this converges to a matrix R, the latter satisfies

$$R = [-Q_0 - R^2 Q_2] Q_1^{-1},$$

which is equivalent to (6.29).

In practice, the iterative procedure is terminated when two successive estimates of R are sufficiently close, according to some criterion. Thus, one can fix a value of ϵ and stop the iterations when

$$\max_{i,j} |R_{n+1}(i,j) - R_n(i,j)| < \epsilon.$$

Experiments show that, for a given ϵ, the number of iterations required to compute R increases with the offered load. The closer the system is to instability, the longer it takes to find a solution. However, this effect can be alleviated considerably by employing a more complicated iterative scheme with 'logarithmic' convergence.

6.3 Generalizations

Consider now a queue in a Markovian environment, where the arrivals and/or departures may occur in batches. As before, the state of the process at time t is described by the pair (X_t, Y_t), where X_t is the state of the environment (the operational mode) and Y_t is the number of jobs in the system. The state space is $\{0, 1, \ldots, N\} \times \{0, 1, \ldots\}$. The variable

6.3 Generalizations

Y_t may jump by arbitrary, but bounded, amounts in either direction. In other words, the allowable transitions are:

(a) From state (i, j) to state (k, j) ($0 \le i, k \le N$; $i \ne k$; $j \ge 0$), with rate $a_j(i, k)$;
(b) From state (i, j) to state $(k, j + s)$ ($0 \le i, k \le N$; $1 \le s \le r_1$; $r_1 \ge 1$), with rate $b_{j,s}(i, k)$;
(c) From state (i, j) to state $(k, j - s)$ ($0 \le i, k \le N$; $1 \le s \le r_2$; $r_2 \ge 1$), with rate $c_{j,s}(i, k)$,

provided of course that the source and destination states are valid.

Obviously, if $r_1 = r_2 = 1$ then this is a Quasi-Birth-and-Death process. Denote by $A_j = [a_j(i, k)]$, $B_{j,s} = [b_{j,s}(i, k)]$ and $C_{j,s} = [c_{j,s}(i, k)]$ the transition rate matrices associated with (a), (b) and (c), respectively. There is a threshold M, such that

$$A_j = A \; ; \; B_{j,s} = B_s \; ; \; C_{j,s} = C_s \; ; \; j \ge M \; . \tag{6.32}$$

Defining again the diagonal matrices D^A, D^{B_s} and D^{C_s}, whose ith diagonal element is equal to the ith row sum of A, B_s and C_s, respectively, the balance equations for $j \ge M + r_1$ can be written in a form analogous to (6.15):

$$\mathbf{v}_j [D^A + \sum_{s=1}^{r_1} D^{B_s} + \sum_{s=1}^{r_2} D^{C_s}] = \sum_{s=1}^{r_1} \mathbf{v}_{j-s} B_s + \mathbf{v}_j A + \sum_{s=1}^{r_2} \mathbf{v}_{j+s} C_s \; . \tag{6.33}$$

Similar equations, involving A_j, $B_{j,s}$ and $C_{j,s}$, together with the corresponding diagonal matrices, can be written for $j \le M + r_1$.

As before, (6.33) can be rewritten as a vector difference equation, this time of order $r = r_1 + r_2$, with constant coefficients:

$$\sum_{l=0}^{r} \mathbf{v}_{j+l} Q_l = \mathbf{0} \; , \; j \ge M \; . \tag{6.34}$$

Here, $Q_l = B_{r_1 - l}$ for $l = 0, 1, \ldots r_1 - 1$,

$$Q_{r_1} = A - D^A - \sum_{s=1}^{r_1} D^{B_s} - \sum_{s=1}^{r_2} D^{C_s} \; ,$$

and $Q_l = C_{l - r_1}$ for $l = r_1 + 1, r_1 + 2, \ldots r_1 + r_2$.

The solution of this equation can be found either by spectral expansion, or in matrix-geometric form.

To apply spectral expansion, one forms the characteristic matrix polynomial

$$Q(x) = \sum_{l=0}^{r} Q_l x^l . \tag{6.35}$$

The normalizable solution of equation (6.34) is of the form

$$\mathbf{v}_j = \sum_{k=1}^{c} \alpha_k \mathbf{u}_k x_k^{j-M} \; ; \; j = M, M+1, \ldots , \tag{6.36}$$

where x_k are the eigenvalues of $Q(x)$ in the interior of the unit disk, $\mathbf{u}_k$ are the corresponding left eigenvectors, and α_k are constants ($k = 1, 2, \ldots, c$). The latter are determined with the aid of the state-dependent balance equations and the normalizing equation.

For computational purposes, the polynomial eigenvalue–eigenvector problem of degree r can be transformed into a linear one. For example, suppose that Q_r is non-singular and multiply (6.34) on the right by Q_r^{-1}. This leads to the problem

$$\mathbf{u} \left[\sum_{l=0}^{r-1} H_l x^l + I x^r \right] = \mathbf{0} , \tag{6.37}$$

where $R_l = Q_l Q_r^{-1}$. Introducing the vectors $\mathbf{y}_l = x^l \mathbf{u}$, $l = 1, 2, \ldots, r-1$, one obtains the equivalent linear form

$$[\mathbf{u}, \mathbf{y}_1, \ldots, \mathbf{y}_{r-1}] \begin{bmatrix} 0 & & & -H_0 \\ I & 0 & & -H_1 \\ & \ddots & \ddots & \vdots \\ & & I & -H_{r-1} \end{bmatrix} = x [\mathbf{u}, \mathbf{y}_1, \ldots, \mathbf{y}_{r-1}] .$$

As in the quadratic case, if Q_r is singular then the linear form can be achieved by an appropriate change of variable.

The matrix-geometric solution is of the form

$$\mathbf{v}_j = \mathbf{v} R^{j-M} \; ; \; j = M, M+1, \ldots , \tag{6.38}$$

where the matrix R is the minimal non-negative solution of the matrix equation of degree r,

$$\sum_{l=0}^{r} R^l Q_l = 0 . \tag{6.39}$$

The vector $\mathbf{v}$ and the other unknown probabilities are obtained from the state-dependent equations.

Equation (6.39) is solved by iteration, in a similar way to (6.29). If Q_1 is non-singular, one can use the iterative scheme

$$R_{n+1} = \left[-Q_0 - \sum_{l=2}^{r} R_n^l Q_l \right] Q_1^{-1}, \qquad (6.40)$$

starting with $R_0 = 0$. The procedure is terminated when two successive estimates of R are sufficiently close to each other.

There are also two classes of models with unbounded queue size jumps that have been examined in the literature. One allows jumps of arbitrary size downward (including down to $j = 0$), but requires that the upward jumps are of size 1 (i.e. $r_1 = 1$, $r_2 = \infty$). Such models are said to be of 'G/M/1 type'. That name comes from an analogy with the Markov chain embedded at the arrival instants of the G/M/1 queue: between two consecutive arrivals, the queue size may decrease by any number of jobs (up to the total number present), but may increase by at most 1.

The models of G/M/1 type have a matrix-geometric solution. However, the matrix equation that needs to be solved in order to find the matrix R is of infinite degree.

The other class of models allows unbounded upward jumps but requires that the downward jumps are of size 1 (i.e. $r_1 = \infty$, $r_2 = 1$). Those models are said to be of 'M/G/1 type', from the analogy with the Markov chain embedded at the departure instants of the M/G/1 queue (see section 5.2). The processes of M/G/1 type do not have a matrix-geometric solution; their analysis is considerably more complicated.

6.4 Literature

A good source for matrix-geometric results, including QBD and G/M/1 type processes, is the book by Neuts [6]. It also contains a chapter on the properties and applications of phase-type distributions. The logarithmic algorithm for computing the matrix R is due to Latouche and Ramaswami [2]. The analysis of M/G/1 type models is covered in Neuts [7]. A bibliography of more than 150 related books and articles can be found in [8].

The spectral expansion solution method was presented in Mitrani and Mitra [5]. The treatment in [3] is similar to the present chapter and includes a discussion of the use of generating functions. Some comparisons between the spectral expansion and the matrix-geometric solutions can be found in Mitrani and Chakka [4], and in Haverkort and Ost [1]. The available evidence suggests that, where both methods are applicable, spectral expansion is faster even if the matrix R is computed by the logarithmic algorithm.

References

1. B.R. Haverkort and A. Ost, "Steady-State Analysis of Infinite Stochastic Petri Nets: Comparing the Spectral Expansion and the Matrix-Geometric Method", *Procs., 7th Int. Workshop on Petri Nets and Performance Models*, San Malo, 1997.
2. G. Latouche and V. Ramaswami, "A Logarithmic Reduction Algorithm for Quasi Birth and Death Processes", *Journal of Applied Probability*, **30**, 650–674, 1993.
3. I. Mitrani, "The Spectral Expansion Solution Method for Markov Processes on Lattice Strips", Chapter 13 in *Advances in Queueing*, (edited by J.H. Dshalalow), CRC Press, 1995.
4. I. Mitrani and R. Chakka, "Spectral Expansion Solution for a Class of Markov Models: Application and Comparison with the Matrix-Geometric Method", *Performance Evaluation*, **23**, 241–260, 1995.
5. I. Mitrani and D. Mitra, "A Spectral Expansion Method for Random Walks on Semi-Infinite Strips", in *Iterative Methods in Linear Algebra* (edited by R. Beauwens and P. de Groen), North-Holland, 1992.
6. M.F. Neuts, *Matrix-Geometric Solutions in Stochastic Models: An Algorithmic Approach*, Dover, 1994.
7. M.F. Neuts, *Structured Stochastic Matrices of M/G/1 Type and Their Applications*, Marcel Dekker, 1989.
8. M.F. Neuts, "Matrix-Analytic Methods in Queueing Theory", Chapter 10 in *Advances in Queueing*, (edited by J.H. Dshalalow), CRC Press, 1995.

Index

σ-algebra, 4
c/ρ rule, 110
z-transform, see generating function, 37

response time, 82
Shortest-Remaining-Processing-Time
  policy (SRPT), 115

absorptions, 184
ALOHA, 69
aperiodic state, 163
arithmetic mean, see sample mean, 34
Arrival rate lemma, 50
arrival theorem, 133, 135, 141
arrivals during a random interval, 67

backward renewal time, 50
balance equations, 165, 170, 181
Bayes formula, 8
Bernoulli trials, 28, 38
binomial distribution, 31
Birth-and-Death queueing models, 186
Borel field, 4
bottleneck node, 142
breakdowns, 205
busy period, 86, 91

Cauchy density, 22
Cauchy–Schwarz inequality, 27
central limit theorem, 41, 43
Chapman–Kolmogorov equations, 179
characteristic function, 41, 43
characteristic matrix polynomial, 211, 218
characterization problem, 116
Chebyshev's inequality, 25, 35
closed networks, 123, 136
Cobham's expressions, 102, 103, 107, 108
coefficient of variation, 83

communication blocking, 207
complement of a set, 3
complete probability formula, 8
Completion rate, 56
conditional density function, 17
conditional probability, 7
conservation law, 108
convolution, 37, 39
correlation coefficient, 26
covariance, 26
CSMA, 69

De Moivre–Laplace theorem, 44
decomposition property of a Poisson
  process, 64, 125
disk storage device, 84
dispersion, 24
distribution function, 11

embedded Markov chain, 180
Engset distribution, 199
entry point, 138, 151
Erlang distribution, 69
Erlang's delay formula, 191
Erlang's loss formula, 192
event, 2
expectation, 20, 23, 24
exponential density, 46
exponential distribution, 55

factorial moment, 38
FCFS, 81
FIFO, 81, 96, 168
first completion time, 57
first moment, see expectation, 24
first passage time, 184
fixed-point equation, 83, 143
forward renewal time, 50

221

Index

G/M/1 system, 74
general normal distribution, 42
generalized eigenvalues, 211
generalized left eigenvectors, 211
generating function, 37, 170
generator, 177, 178, 180
geometric distribution, 28, 39
Gittins index, 113
Gordon–Newell theorem, 137

head-of-line policy, see non-preemptive policy, 100
hyperexponential distribution, 119, 209

identity matrix, 126
independent delay system, 79, 97, 129, 136, 193
index scheduling policy, 113
infinitely many servers, 78
instantaneous transition rates, 173, 176, 201
intersection of sets, 3
irreducible Markov chain, 162

Jackson's theorem, 128, 132, 138
joint distribution function, 15

Kleinrock's conservation law, 109
Kolmogorov backward differential equations, 179
Kolmogorov forward differential equations, 179

Laplace transform, 40, 59, 91, 170
law of large numbers, strong, 34
law of large numbers, weak, 34
LIFO, 92
LIFO-PR, 92, 94, 96, 98
limiting probabilities, 162
linearization, 213, 214
Little's theorem, 75, 89
logarithmic convergence, 216
lumping job types, 105

M/D/1, 94
M/G/1, 74, 81, 94, 168
M/M/∞ systems, 192
M/M/k_i, 129
M/M/n model, 188, 191
M/M/n/n systems, 192
M/M/1 queue, 88, 94
M/M/1/N queue, 194
M/M/2 queue, 189
M/M/$n/\cdot/K$ model, 195
M/PH/1 queue, 210
machine maintenance, 79, 197
manufacturing blocking, 207

marginal distribution, 15, 129, 203
Markov chain, 157
Markov inequality, 27
Markov process, 157, 172
Markov property, 156, 172
Markov-Modulated Poisson Process, 204
matrix-geometric solution, 215, 218
mean value analysis, 138
measure, 4–6
memoryless property, 29, 55, 60
MMPP, see Markov-Modulated Poisson Process, 204
modified geometric distribution, 31, 89, 172
moment, 24
multiclass networks, 146
Multiple modular redundancy, 31
multiprogrammed computer, 145

negative binomial distribution, 36
negative customers, 153
neural networks, 153
non-preemptive policy, 100
normal approximation, 62
normal distribution, 41
normalizing constant, 137, 145
normalizing equation, 165, 181
number of visits, 130, 139

offered load, 77
one-step transition probabilities, 157
open network, 123, 124
operational modes, 201
order statistics, 18, 32

paradox of residual life, 53
passage times, 130
PASTA property, 65, 82, 89, 169
pdf, see probability density function, 13
periodic state, 163
PH, see phase-type distributions, 209
PH/PH/n queue, 210
phase-type (PH) distributions, 208, 209
Poisson distribution, 45
Poisson process, 59
Poisson process as a limit, 66
Pollaczek–Khinchin formulae, 83, 171
preemptive-repeat policy, 101
preemptive-resume policy, 100, 104
priority scheduling, 99
probability, 4
probability density function, 13
Processor-Sharing, 94, 97, 98, 115
PS, see Processor-Sharing, 94

Index

QBD, see Quasi-Birth-and-Death process
Quasi-Birth-and-Death (QBD) process, 202, 204
queue in Markovian environment, 201
queueing network, 122

random experiment, 1
random modification, 50
random observation point, 50
random process, 156
random variables, 10
random walk, 168
reachable state, 162
recurrent non-null state, 164
recurrent null state, 164
recurrent state, 162
renewal equation, 49
renewal function, 49
renewal intervals, 47
renewal process, 47
repairs, 205
residual life, 50
response time, 130
response time law, 80
Round-Robin policy, 94
routeing, 122
routeing probabilities, 124

sample mean, 34
sample points, 2
sample space, 1
scheduling policy, 92
service quantum, 113
shortest interval, 177
Shortest-Expected-Processing-Time-first (SEPT) policy, 112
Shortest-Processing-Time-first (SPT) policy, 103
software reliability growth, 34
spectral expansion, 211, 212, 218
spectral methods, 179

SPT, 107
squared coefficient of variation, 26
SRPT policy, 115
standard deviation, 26
state diagram, 158, 174
state space, 157
steady-state, 75, 161, 180
steady-state distribution, 164, 181, 203
steady-state theorem, 164, 181
stochastic process, 156
strong law of large numbers, 34
superposition property of a Poisson process, 62
symmetric scheduling policies, 96, 147

telephone network, 175
throughput, 138
time to absorption, 184
time-homogeneous, 158, 172
total average service, 131
traffic equations, 125, 131, 137, 147, 150
transaction system, 79
transient distribution, 178
transient state, 162, 167
transition probability functions, 173
transition probability matrix, 158

uniform density, 15
union of sets, 3
union property of a Poisson process, 62
utilization, 77, 142
utilization law, 78

variance, 24
vector difference equation, 211
virtual load, 108

waiting time, 82
weak law of large numbers, 34
window flow control, 143